Dr. Eleanor's
BOOK OF
Common Ants OF
Chicago

DR. ELEANOR'S

BOOK OF

Common Ants

OF

CHICAGO

ELEANOR SPICER RICE,
ALEX WILD, *and* ROB DUNN

THE UNIVERSITY OF CHICAGO PRESS
Chicago and London

The University of Chicago Press, Chicago 60637
The University of Chicago Press, Ltd., London
© 2017 by Eleanor Spicer Rice, Alexander Wild, and Robert Dunn

Published 2017

Printed in Canada

26 25 24 23 22 21 20 19 18 17   1 2 3 4 5

ISBN-13: 978-0-226-26680-0 (paper)
ISBN-13: 978-0-226-26694-7 (e-book)
DOI: 10.7208/chicago/9780226266947.001.0001

LCCN: 2017026799

♾ This paper meets the requirements of ANSI/NISO Z39.48–1992
(Permanence of Paper).

CONTENTS

Preface, vii

Introduction, 1

PREFACE

The Chicago Queen of Ants

In the story of the societies of Chicago, Mary Talbot is the undisputed queen. In the 1920s, when Al Capone and rival gangsters fought to control what was going on above ground in the windy city, Talbot crouched down in order to understand the more sophisticated societies underfoot, those of the ants.

Mary Talbot grew up playing outside. The story of her childhood is like that of many biologists (including both Dr. Eleanor and Dr. Moreau). She gathered things. Snails. Insect parts. Snakes. And she studied them, sometimes with a poke and a prod, more often with patience. She took notes on what she found. Her older brother led the way when she was very young, but soon she was on her own, out among the weeds of her rural town, encountering life.

Many children grow up wandering along trails, chasing mysteries. But most abandon such searches for the more mundane quests of adulthood. Not Mary. She would be, she thought from an early age, a scientist, someone who would spend her life making discoveries. Mary Talbot grew up in a time when very few women were scientists of any kind. Fewer still were entomologists, scientists who study insects. Yet at a relatively early age Talbot was sure that this was just what she wanted to do. Undoubtedly she encountered barriers, and if more were recorded about Talbot, we might know something of their nature. What we do know about are Mary's discoveries, truths of nature revealed in the woods.

From among the many creatures that interested her, Talbot chose to focus on insects as an undergraduate student at Denison College (in Granville, Ohio). She went on to a master's degree, during which time she studied a beetle that lives only with ants, hiding among them, in plain view. This is what Talbot also came to want, to hide among the ants, watching them without being noticed, to know their ways. Talbot's next step was to earn a doctoral degree, and there seemed to be only one place worth doing so, one place where the colonies of social insects were being taken as seriously as the cities of humans, Chicago. Talbot wanted to study in the lab of A. E. Emerson. If she was to become the queen of ants, he was the king of termites.

Emerson lived in Chicago and worked at the University of Chicago. Chicago is an unusual place to study termites. Just one species, *Reticulitermes flavipes*, lives in the city and it is neither common nor easy to study. But as a young man Emerson traveled to the tropics to meet with the great explorer William Beebe. (Beebe's expeditions knew no bounds. He explored clouds, jungles and even, inside a spherical submarine of sorts, the deep sea.) Emerson, upon seeing the tropics with Beebe, fell in love, both with the forests themselves and with their termites. He went on to name hundreds of new species of termites and to study their cities, their elaborate empires, and how they had evolved. In considering these empires, he dissected them in the way that one might dissect a body, looking at the parts in order to understand the whole. He saw termite mounds that were 20 feet tall, others that went 20 feet into the ground, others still in which queens larger than his thumb laid hundreds of thousands, maybe millions of eggs, eggs that would rise into an empire. The size of the constructions the termites built fascinated Emerson. The termites had to make all of the decisions that the people who built Chicago made. Decisions about where roads went, where babies live, how to move food, how to support a building in a way that it would not collapse. The termites made all of these decisions without ever having a conscious thought. The termites were dumb and yet, somehow, also, sophisticated, more sophisticated it probably seemed than the people running his university, or city (though maybe I am projecting).

The more Emerson looked at the termite colonies, the more the nests themselves seemed to him like organisms in their own right, composed of individual termites rather than cells. A human body has cells that play different roles—eye cells, heart cells, and blood cells, for instance. In the same way, different termites do different jobs. There exists division of labor. The signals, too, among termites are like the signals among neurons. The food moving among termites are like nutrients moving in the blood. In considering this sophistication, Emerson found himself at a crossroads. Only one conclusion seemed possible. The termite mounds were not made up of individual termites at all but really they were more like bodies, the entire termite mound analogous to an entire human body. The mound was an organism, and just as the electrical firing in our individual (and unsophisticated) brain cells leads to complex decisions inside our heads, the individual decisions and actions of tiny termites lead to sophisticated decisions inside the termite mound. When contemplating the termites out in the savanna, Emerson sometimes felt as though he had encountered organisms so sophisticated that they were studying him back.

Talbot knew about Emerson. She wanted to work with him, to understand the societies of insects. And so it was that she landed in Chicago. But upon arriving in Chicago, rather than being drawn into his studies of termites, she began considering something else. As she walked around the city, she saw so many kinds of societies all around her, the societies of ants she had studied in Ohio. They were like old friends, superorganisms with which she would like to spend some time. Superficially, Talbot lived in the society of humans, Chicago's roaring, gangland twenties, but during her years in Chicago she spent far more time among the societies of ants.

In and around Chicago Talbot found ant species wherever she

looked. In patches of forests, clumps of grass, and even piles of sand. She found 90 species in total. Some lived in acorns. Others in giant thatch mounds. Others in trees. Talbot was clear to note in her work that most of the species she encountered had not been studied. Talbot wanted to study them, in order to understand all the different ways to be a society.

In Talbot's time, societies of ants were studied as part of basic biology, an attempt to understand how the living world around us works. They still are, but something new has happened. The study of ants has become useful.

In the last decade, around the world, biologists have made many discoveries by considering the ways in which insect societies deal with the challenges our bodies (or our cities) have to deal with. They have found new antibiotics by studying the compounds ants use to keep themselves healthy. They have found new ways to build roads by studying the ways ants make their paths and discovered new ways to model the transfer of computer data by watching how ants transfer, mouth to mouth or through the air, their own signals. They have better understood human longevity by studying how and why ant queens live so much longer (many years) than do workers.

Many of the recent and useful discoveries made by studying ants build on Talbot's work. Talbot studied the societies of more species of ants than almost any other ant biologist, living or dead, but not in Chicago. Talbot did most of her work after she left Chicago. Talbot moved to Linwood College in St. Louis, Missouri, where she taught for several decades and then, in the summers, did her work on ants at the E. S. George Reserve in Michigan, not far from where I grew up. Talbot pioneered research on winter ants, those that store honey in their butts for the lean times. She studied, more cleverly and comprehensively than anyone before or since, when queens and fragile males (male ants do little more than mate) fly toward each other. She resolved what winnow ants do in the winter. All of this work required heroic digging and determination of which few biologists have been capable since. In excavating winter ants, Talbot had to dig holes tens of feet deep, with a shovel, on her own, to get to the queen. She could not have been happier.

What Talbot did in Chicago was simpler than this work. What she did was to reveal the 90 ants that she knew, upon more detailed study, would reveal more truths to us, about them and even about us. Talbot revealed the Chicago ants and then left them to whoever else might step in, but no one did. In the 90 years since Talbot studied the ants of Chicago, only one other study of the windy city's ants has been published, and that study mostly just reviewed what Talbot had already found. Most of what might be discovered about the ants of Chicago remains to be discovered.

But that is changing. The first effort at this discovery has already begun to happen. It is represented in this book, in the stories of the ants of Chicago as revealed by samples taken by those who are part of a project called the School of Ants. The School of Ants aims to give kids and adults around the United States (and now in parts of Italy and Australia) the wherewithal to go into their backyards and collect ants in order to document where ants of different species live. The project is new, but already the discoveries have been big. Whether while working on their own, with Dr. Corrie Moreau at the Chicago Field Museum, or with Sean Menke at Lake Forest University, these young ant scientists have begun to give us the most detailed look at the ants of Chicago since the 1944 work of Mary Talbot.

The catch is that in order to make discoveries, one needs to know what is already known, where the last path ended and where a new one might begin. Before now, there has been no book describing what we already know about the common ants of Chicago. Here Dr. Eleanor, an ant biologist, tells their stories. These stories are fun, but they are also something more; they are a clear indication of where the paths end in our understanding of these species With *Dr. Eleanor's Guide to the Common Ants of Chicago* as a launching point, anyone with enough interest, patience, and perseverance might discover something about the ants living in and under Chicago. Those ants have things to teach us about running a civilization and, to the extent that each colony really is a kind of superorganism, how individuals can work together, obeying simple rules, to create something far greater than any of them could imagine on their own.

Rob Dunn

INTRODUCTION: WHAT'S THE BIG DEAL ABOUT ANTS?

Before you dive into the stories of individual ant species, let's start with some basic ant biology and a little natural history.

Ants surround us, occupying nearly every type of habitable nook and cranny across the globe. Right now, ants snuggle up to your house, lay out their doormats in front of the trees in your yard, and snooze under park benches. Some even nest inside the acorns littering the ground!

We might not always notice them, but they're there, and they shape, literally shape, our world. Look at the colossal trees in our

forests. Ants like winnow ants plant the forest understory, ultimately contouring the distribution of plants in the undergrowth and even of those giant trees. Other ants help turn soil (more than earthworms in some places!), break up decomposing wood and animals, and keep the tree canopy healthy by providing rich soil nutrients for saplings, protecting tender leaves from predators, and regulating tree pest populations.

Ants creep across our yards taking care of business for us in much the same way. They eat termites and chase caterpillars out of our gardens. Even though some people think of ants as the tiny creatures that ruin their picnics, of the nearly 1,000 ant species living in North America, fewer than 30 are true pests, and fewer still actually can hurt us. Most ants spend their time pulling threads, stitching together the quilt of the natural world. Without these threads, the quilt would fall apart, becoming disconnected pieces of fabric.

In this book, you will meet our most common ants. Odds are you can see these ladies tiptoeing all around you. See how beautiful they are, with their spines and ridges, their colors and proud legs, each feature lending itself to the individual's task. See their work, how they help build the world around us as they move about our lives.

What's in an Ant?

Like all insects, adult ants have three body segments: the head, the thorax, and the abdomen.

Heads Up

Windows to the world, ant heads are packed with everything ants need to interact with their environments. With tiny eyes for detecting light, color, and shadow; brains for memory and decisions; mouths for tasting; antennae for touching and smelling, ant heads are one-stop shops for sensory overload.

petiole

thorax

antennae

gaster

Thorax

Ant thoraxes are mainly for moving. Although an ant's nerve cord, esophagus, and main artery course through its thorax, connecting head to bottom, thoraxes are mostly legs and muscle. Every one of an ant's six legs sticks out of her thorax, and when queens and males have wings, those wings stick out of the top of the thorax.

Abdomen and Petiole: One Lump or Two?

The abdomen is where all the action happens. This ant segment holds all the critical organs. Almost all of an ant's digestive system is packed into its booty, as well as tons of chemical-emitting glands, stingers and trail markers, the entire reproductive system, and most of its stored fat. Many ant species also have a crop, which is a special stomach in their abdomens that does not digest food. Ants use this stomach as a backpack to carry food back to the nest. There they share the food with their sisters by vomiting it back up and then spitting it into their sisters' mouths.

The first part of an ant abdomen is called the petiole. The petiole is that really skinny section between the thorax and an ant's big fat bottom, which we often call a gaster. The petiole gives ants their skinny waists and flexibility when they move around. A lot of people interested in identifying ants check the petiole first to see if it has one bump or two as a first step in determining the species.

Where's the Nose?

Unlike us, ants don't have noses. Instead, they smell and breathe with different body parts. To smell, they mostly use their antennae. To breathe, ants have little holes all along their body called spiracles, which they can open and close. When the spiracles open, air rushes into beautiful silvery tubes that lace the ants' insides and bathe their organs with the oxygen they need to survive.

The Ant Life Cycle

Like butterflies, beetles, and flies, ants grow up in four stages: a tiny egg, a worm-like larva, a pupa, and the adult stage that most of us recognize as ants.

After hatching, larvae molt several times.

Eggs

For most species, only the queen lays eggs that become workers. Most of the time, eggs are creamy-colored orbs smaller than the period at the end of this sentence. When queens fertilize the eggs, females hatch. When they don't, males hatch. Sometimes, like when the colony is just getting started, queens lay eggs called trophic eggs. Delicious and nutritious, workers and developing larvae eat trophic eggs and use that energy to help the colony grow. Occasionally, workers lay eggs, but queens and other workers sniff about the nest and gobble those eggs as soon as they find them, no bacon needed.

Larvae

The only time an ant grows is when it's in the larval stage. Most ant species' larvae look like wrinkly grains of rice or chubby little maggots. Fat pearlescent white tubes with plump folds and wormy mouths, ant larvae are the chicken nuggets of the insect world. When colonies get assaulted by other

Workers tend eggs and larvae.

ants or insects, these little chub monsters are usually the first to go. Because they have no legs, they can't run away, and they make easy meals for anyone able to break into the nest.

Unable to feed themselves, larvae sit like baby birds with their little mouths open, begging for workers to spit food down their gullets. When workers give larvae special food at just the right time, their body chemistry changes and they grow up to be queens. As larvae grow, their skin gets tight on their bodies. They wiggle out of the tight skin as if shedding an old pair of blue jeans, revealing a brand new, bigger skin underneath. Never wasting the chance for a snack, workers squeeze whatever liquid remains from those discarded larvae skins and sometimes feed the solids back to larvae.

If you look at older larvae under a microscope, you'll see sparse hairs jutting out of their supple flesh. I know a scientist who wanted to find out why they have these luxurious locks, so he gave larvae haircuts and watched what happened. The verdict? Shorn larvae fall over like little drunk sailors. They need their hair to anchor them to surfaces.

Some ant species, like this one, spin silken cocoons.

Pupae

When larvae grow big enough, they quit eating and get really still. The larvae of some ant species will spin a silken cocoon around themselves to have a little privacy during this pupa stage; others will just let it all hang out. Although they don't look like they're doing much during this stage, their bodies are changing and shifting around inside that last larval skin. They're developing legs and body segments, antennae and new mouthparts. They're turning into the ants we recognize.

When they're ready, they squeeze out of their last skins, emerging as full-grown ants. At first, they tumble about like baby deer, unsure on their legs, soft and pale like the larvae they used to be. But after a while, their skins darken and harden, their step becomes surer, and they begin their work as adults.

Workers

The two most important things to remember about ant workers are the following: First, all workers are adult ants. Once an ant completes its metamorphosis (when it is recognizable as an ant), it will never grow again. When you see a little ant, it's not a baby ant; it's just a species of ant that is really small. Second, all workers are female. Workers do nearly every job, so pretty much every ant you see walking around is a girl. While queens get the colony rolling and keep it strong by laying eggs, workers get the groceries, keep intruders out, take out the trash, feed the babies, repair the house, and more. When we talk about ant behavior and the special characteristics of ants in this book, we're talking about the behaviors workers exhibit in the natural world, since they are the colony's only contact with the outside world.

Queens

Despite their regal moniker, ant queens are mostly just egg-laying machines. When queens first emerge, they usually have wings, but after they find a lucky someone (or someones) and mate, they rub off their wings, let their booties expand with developing eggs, and go to town eating food and popping out eggs. Protected deep within the nest, workers feed queens and keep life peachy for them so they can produce healthy eggs for the colony.

An ant queen, with her large gaster, is ready for egg laying.

Males

Male ants are easy to discount because they don't seem to do too much around the colony. Unlike their more industrious sisters, male ants refrain from cleaning up around the house, taking care of the babies, going out to get food, or keeping bad guys out. The one thing male ants do for the colony is mate with queens.

To date, scientists have spent very little time studying male ants. But these mysterious and weird-looking creatures invite a closer look. Compared to their sisters, most male ants have tiny heads and huge eyes. Often, they look like wasps. Nobody knows for sure what the boys do when they leave the nest. What are they eating? Where do they sleep? Why doesn't anybody seem to care? As you read this, researchers are trying to finally put an end to our male ant ignorance.

What's in an Ant Colony?

Many different types of ants will nest in pretty much any type of shelter. While big headed ants push up their earthen mounds for all to see, acrobat ants might have their mail delivered to a tiny piece of bark on a tree limb, and winter ants scurry down inconspicuous holes in the ground to their underworld mansions.

While ant nests differ greatly, when you crack one open, you'll most likely find lots of workers (the ants we most often see in the "real world"), a queen (many species have several queens), and a white pile of eggs and larvae.

Most ants carry out the trash and their dead, piling them in their own ant dumps/graveyards called middens. A good detective can learn many things from going through someone's trash. If you examine a midden, you can get a good idea of what the ants have been eating and whether or not the ants are sick or at war with other ants. You'll probably discover bits of seeds and insect head capsules stuffed in with the dead ants. When tremendous numbers of dead ants litter the piles, it's likely the colony is sick or warring with other ants.

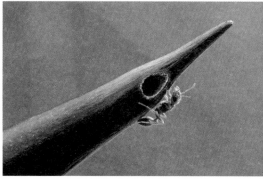

Back inside the nest, the ants busy themselves with their daily anty lives. You can take some cookie crumbs and call them out to you. See how they sniff the earth with their antennae, each one a living being experiencing the world, doing its special job. Watch them communicate, following one another under blades of grass and around pebbles, stopping every now and again to touch one another's faces, clean their legs, investigate their surroundings.

Ants nest in the ground, in thorns, and anywhere else.

Ants saturate our environment, from our homes to the sidewalks, city streets, and forests spread all around us. They are our neighbors, our friendly fellow citizens working away as we work. It's time we introduce ourselves.

01 HOBBIT ANT

SPECIES NAME: *Stenamma* spp.

SIZE: 5.08 mm (0.2 in.)

WHERE IT LIVES: *Stenamma* ants nests in small, discrete cavities under stones, in hollowed twigs, or empty nuts. Because of their cryptic lifestyle, scientists aren't sure of their exact whereabouts throughout the United States, but we know they live in Chicago and the surrounding area.

WHAT IT EATS: Any tiny soil invertebrate will do for the *Stenamma* ant, who will hunt and scavenge her favorite meals.

When considering fictional characters you identify with, you might choose a scrappy moral heroine like Scout Finch from *To Kill a Mockingbird* or maybe the clever and sleuthy Sherlock Holmes. Or you

could be more the action hero type, troubled-yet-tender like Wolverine or charming and unflappable like Iron Man. Me? I'm more of a hobbit at heart. Any hobbit. Outwardly, I may not look like Tolkien's furry-footed minikins, but we have many things in common. For starters, hobbits and I are Most Concerned about when we're going to eat next, and we always hope it's going to be good. We like going barefoot. We prefer curling up in our comfortable homes with our friends and family to Batman-like action-packed adventures. *Stenamma* ants are the hobbits of the ant world. That's why I like them so much. Like hobbits, they're small and furry. But also like hobbits, we should never make the mistake of misjudging them.

Because they have no common name, just for this little while, for fun, let's call them hobbit ants. But before I tell you about hobbit ants, you need to know about an ant hero named Mary Talbot, one of the greatest ant experts, or myrmecologists, who ever lived. She was born in Columbus, Ohio, in 1903 and, like me, spent her childhood following her big brother around, helping him lift logs, beat grass, peel bark, and dig holes to find the living things that surround us. When she grew up, she was still lifting, beating, peeling, and digging. She was patient with the earth, and careful, and she unearthed thousands of ants and thousands of discoveries—new species, who lives where, how some ants mate, how they share this crawling world with each other, and how their balance shifts and flips and tugs over time.

In the 1930s, Mary picked apart Chicago, unknotting roots, sifting soil, looking for the ant networks that course through the city's secret spots like traffic in the streets. While working in oak maple woods near Lake Michigan, Mary said she saw the "five-leafed ivy and wild grape" vines. She walked over ground "covered with many types of spring flowers" and noted the ancient "hemlock, a remnant from glacial periods," stretching its furry fringes up toward the sky and out over the forest floor. Then she found them, our hobbit ants, hiding beneath the trees.

Hobbit ants, like their new namesake, are diminutive even by ant standards. Measuring about a tenth of an inch, their workers could turn themselves around on the flat head of a straight pin. Two species make the most common list, *Stenamma diecki* and *S. brevicorne*, but several more can be found across the United States.

When Tolkien describes a hobbit home, he says, "In a hole in the ground there lived a hobbit. Not a nasty, dirty, wet hole, filled with the ends of worms and an oozy smell, nor yet a dry, bare, sandy hole with nothing in it to sit down on or to eat: it was a hobbit-hole, and that means comfort."

Hobbit ants also live in their own comfortable hobbit holes. All across the United States, northern Mexico, and Canada in woodlands, grasslands, and meadows, they excavate small, inconspicuous cavities under stones or leaf litter, in twigs' hollow holes, or in empty nuts. Sometimes, you can crack open an ordinary-looking acorn only to find it's actually a clean, dry hobbit ant mansion, with eggs and larvae hanging from the sides like pearlescent chandeliers, and a dozen to a couple hundred workers eagerly tending to them and fluffing up their one queen, taking care of daily hobbit ant business). And those queens! Like hobbits, these miniature wonders can live long, long lives. Brave little hobbit ant queens can live up to 18 years.

And much as the hobbit Bilbo Baggins said of his kind, "We are plain quiet folk, and I have no use for adventures. Nasty, disturbing, and uncomfortable things," the hobbit ant queen shies away from adventure and helps her colony avoid predators by adopting a cryptic lifestyle, choosing those inconspicuous spots to set up her hobbit hole. Her timid workers expand on those cryptic habits, preferring to be unobtrusive whenever they can, moving slowly, sometimes operating in winter, when most ant monsters sleep. When they do encounter danger, instead of gnashing mandibles or ramming exoskeletons, hobbit ants often play dead.

Hobbit ants benefit from and even thrive in their timidity, gobbling up teeny invertebrates scuttling through the soil and taking

Hobbit ants are friendly, tiny, and elusive.

life at their own pace. While many ant species make babies every year, hobbit ants make babies only when times are good. With such a long-lived queen, it's no biggie to skip a year or two until they have enough resources for new family members.

Occasionally, our little hobbits summer away from home. In warm weather they split up the colony into smaller nests in several acorns or twig cavities. With satellite nests like these, hobbit ants can cover more foraging area while the weather's hot. In the colder months, workers reconvene in a huddle, as everybody comes home to roost together.

Mary Talbot found hobbit ants in one of the several locations she sampled across Chicago. That doesn't mean that hobbit ants weren't scooting around other places. It can be hard to predict a hobbit ant habitat. Sometimes they abound in one area while just a few feet away you won't find hide nor tiny, tiny hair.

Mary knew what *The Hobbit*'s Thorin knows: "There's nothing like looking, if you want to find something." Mary looked. Persistent and unwearied, in love with ants and the world that held them, she looked. There, under the primordial hemlocks reaching out to windy Lake Michigan, she found a hobbit hole or two and saw the tentative antennations of hobbit ants questioningly poking between Earth's fibers. Their great great grandchildren still question today from their own hobbit holes, holes you might find if you get down low enough and slow enough and look.

02 CARPENTER ANT

SPECIES NAME: *Camponotus pennsylvanicus*

AKA: black carpenter ant

SIZE: 7–13 mm (0.25–0.51 in.)

WHERE IT LIVES: Black carpenter ants prefer to nest in living, standing trees but will also nest in logs and wood in human structures. They are widespread in the Northeast and Mid-Atlantic regions and are found throughout the Southeast (with the exception of Florida) and into the eastern Midwest region.

WHAT IT EATS: Omnivores, black carpenter ants eat protein foods, including other insects, as well as sugary foods, like the honeydew produced by aphids.

The black carpenter ant is one of the United States' largest and friendliest ants. Because of their size and pleasant disposition, they make excellent ambassadors between the ant and human worlds.

You can identify black carpenter ants by their size (BIG) and the light dusting of golden hairs on their head and thorax that settles on their abdomens. Unlike workers in some other ant species, black carpenter ant workers vary in size and shape within the colony. Between a quarter-inch and a little over a half-inch long, a small black carpenter ant can comfortably stretch over a plain M&M, and a large one can just about reach across a dime. Colonies have between about 350 to almost 2,000 workers, which, depending on worker size, works out to be almost 200 dollars' worth of dimes banging around inside those trees or, more deliciously, up to 40 bags of M&M's.

Breakfast for Ants

When I was little, I would take my breakfast crumbs out to my front yard to feed the black carpenter ants living in the willow oak trees. I would build little piles of bacon and toast for them on top of oak leaves and wait for them to lumber out from holes hidden in the bark at the bases of the trees.

I loved those ants. I was fascinated by the way they walked around like miniature black horses, exploring their way with their elbowed antennae, stopping every now and then to gently tap their sisters and give each other waxy kisses. If I pressed my ear against the tree near their entryway, I could hear them crackling about their business inside. If I sat still, they would come up to my hands and gingerly pick crumbs off my fingers. If I picked one up, she would explore my arm and shirt. If I squeezed her, she would give me a pinch with her tiny jaws. It never hurt.

They're called carpenter ants because they are particularly good at woodworking. They like to nest in living, standing trees using their sturdy mandibles to excavate tunnels and rooms in dead limbs or in dead wood in the tree's center. Many people see black carpenter ants living in their trees and think the ants are killing the trees. But black carpenter ants actually have a history of helping trees. They have an appetite for tree pests like red oak borers, and they spend a lot of their time foraging around their home, plucking pests off the bark. The trees housing my carpenter ants 25 years ago are still standing today.

Because of these woodworking skills, some people see carpenter ants as household nuisances. While black carpenter ants can make

their tunnels in the wood of people's homes, they often point homeowners to bigger problems: damp and rotting wood from a leak or drip or other pests living in that wood. When wood becomes soaked through, carpenter ants can easily use their jaws to snap it away and bore their tunnels. If homeowners keep their wood dry, carpenter ants will usually stick to the trees. That is, unless the homeowners have pests like termites or wood beetles snacking away inside their walls.

Sometimes black carpenter ants will happen upon such a treasure trove of food and set up camp right next to their grocery store. Can you blame them? Haven't you ever dreamed of living next to your favorite doughnut shop or fried chicken restaurant? Instead of attacking carpenter ants for living in your walls, use them as helpful guides to identify the real problem.

Ant Speak: Decoded

I used to think my carpenter ants might like some of my bologna sandwiches from lunch, but I couldn't get as many takers at lunch-time as I got early in the morning. That's because carpenter ants are mostly night owls, foraging from dusk until dawn. Black carpenter ants have pretty good vision for ants, and they use that vision to help them take shortcuts from their house to food in the early morning and when the moon is out.

When they aren't following their sisters' chemical trails, they remember landmarks like pebbles and sticks to help them find their way home. These landmarks save time for black carpenter ants, who can sometimes forage up to 100 yards from their nest. That's the human equivalent of walking more than 11 miles for food. On new moon nights, when it is totally dark, black carpenter ants take no shortcuts and feel their way through the night, keeping their bodies close to structures.

When carpenter ants find food, they run back to the nest, laying a chemical trail behind them. Once inside the nest, they do an "I found something awesome" dance to get their sisters awake and excited enough to follow them. The hungrier the ants, the more vigorous the dance. The excited sisters then rush out of the nest in search of the chemical trail that will lead them to the food. Carpenter ants, like many other ant species, have little built-in knapsacks called crops inside their bodies. They fill these crops with liquid food to take back home. When they meet their sisters on the trail, they stop and have a little conversation that goes something like this:

ANT HEADING OUT TO FOOD: "Hey, what's up?"
ANT RETURNING FROM FOOD: "Are we from the same nest?" (They check

this by tapping each other on the head with their antennae to see if they smell alike.)

HEADED-OUT ANT: "Yeah, but I'm not sure what I'm even doing here. I'm just following this trail." (She moves her tapping antennae closer to her sister's mouth.)

RETURNING ANT: "Oh, wow! I should have told you earlier. Some kid spilled his Dr. Pepper down the street, and it is DELICIOUS. Everybody's over there now drinking it up. Want to try?"

HEADED-OUT ANT: "That sounds awesome. Of course."

Returning ant spits a little droplet from her crop into headed-out ant's mouth. Headed-out ant drinks it and agrees it is awesome. Awesome enough, in fact, to continue running down the trail.

When I was a child, I saw black carpenter ants having these sorts of conversations all the time and thought they were kissing. When I grew up, I learned that I already knew much about black carpenter ants from watching them as a child. Their colony size, where they nest, and how they eat have all been scientifically dissected and explored as thoroughly as the ants themselves explore the dark tunnels of their homes. Scientific papers explain how they talk to each other, when they're awake, and why they don't want bologna on hot summer afternoons. Every delicate golden hair on the black carpenter ant's rump has been counted and cataloged. These discoveries took many decades to document. All of them can be remade any morning by each one of us, holding our breakfast crumbs, waiting patiently in our front yards.

03 ODOROUS HOUSE ANT

SPECIES NAME: *Tapinoma sessile*

SIZE (WORKERS): 2.25–3.2 mm (0.09–0.13 in.)

WHERE IT LIVES: Odorous house ants nest indoors (under sinks and doormats and in insulation and dishwashers) and outdoors (under rocks and in garbage cans, potted plants and exposed soil). They are heavily concentrated along the West Coast and in the Southwest region; found in the eastern Midwest states; and sprinkled along Florida and the Northeast and Mid-Atlantic regions.

WHAT IT EATS: Odorous house ants eat honeydew, a sugary liquid made by small, sap-feeding insects like aphids and scales, and other sugary food left out by humans. They also eat dead insects and spiders.

People across the United States call me all the time to tell me they have ants in their houses. It's one of my favorite parts of knowing a little bit about insects. From my grandmother Ina down in Opelika, Alabama, to my good friend Ariana out in Los Angeles to my friend Sarah's grandmother's friend up in Baltimore, the call is always the same: "Help me! I'm under attack! I've got ants in my kitchen!"

I love these calls because they make me feel like a real live wizard. Here's why: Across the United States, there are only three or four types of ants that often wander into people's kitchens. By asking a few questions, I can usually narrow the identity of the particular trespassers down to the species through a process of elimination. It's simple, but it seems like magic to the people who are calling. To let you in on the secrets of my sorcery, here is the phone conversation I had with Sarah's grandmother's friend (SGF):

SGF: "Help me! I'm under attack! I've got ants in my kitchen!"
ME: "Are they big or little?"
SGF: "They're tiny!"

Clue 1: They are tiny. Now I know she doesn't have big carpenter ants or the less probable field ants. She also doesn't have Asian needle ants.

ME: "What color are they?"
SGF: "I gotta look at them? Hold on. I gotta get my reading glasses. Hold . . . on . . . OK! They're black!"

Clue 2: They are black. So, Sarah's grandmother's friend doesn't have pharaoh ants or fire ants. Plus, she probably doesn't have the brown Argentine ants. One more answer and I'll know what she has in her kitchen. Time for my big finish.

ME: "Here's what I want you to do. I want you to squish one. I want you to roll it between your fingers and put it up to your nose and sniff it."
SGF: "I'm sorry, what?"
ME: "Just do it. Tell me what it smells like."

Sarah's grandmother's friend squishes. She makes the I'm-squishing-an-ant sound people make, which comes out as a mix between "ooh!"

(fun!) and "eew" (gross). The result of this squish-and-sniff will tell me whether she has little black ants (about half the size of a sesame seed) or odorous house ants (a little bigger than a sesame seed).

SGF: "It smells . . . it smells good! It smells!"

Clincher: They have an odor. Like most people with ants, Sarah's grandmother's friend has odorous house ants partying in her kitchen. Their telltale smell gives them away. She's a lucky lady. Neither dirty nor dangerous, this top home pest—also known as the sugar ant—can provide hours of entertainment for anyone willing to share space with them. Follow them home to see how they bunk! Put out food and see how long it takes them to find it! Lay an *E.T.*-style trail of snacks to shift their ant highways! Possibilities for fun abound.

Country Ant, City Ant

Unlike some of the ant species that pester people around the country (crazy ants or Argentine ants, for example), odorous house ants did not migrate here. They are US natives. Named for a defensive odor they emit from their rumps that some describe as "spoiled coconut suntan lotion," they nest in natural environments like the woods or

in pretty much any manmade locale like potted plants, under door-mats, or in cars. As with Aesop's country mouse and city mouse, "country" odorous house ants (those living in natural, wooded areas) and "city" odorous house ants (those living in manmade environments) lead different lifestyles.

In the country, odorous house ants play an important role keeping the earth a clean, green machine. They work in concert with other forest bugs to keep tree canopies healthy and ensure a proper ecological balance with plenty of species hanging around. They also help accelerate decomposition and promote nutrient flow by eating dead insects and animals and nesting in and under rotting wood, in acorns, and in abandoned insect homes.

Yes, out in the country, they live the quiet life and have small colonies of a few hundred to a couple thousand workers. But once they move into cities, odorous house ants go a little wild. Their populations explode, sometimes spanning entire city blocks, and they blanket lawns and kitchen counters with greedy scouts sniffing around for a sugar fix.

When we build cities, we also build the perfect environment for odorous house ants to go berserk. First, it's easy for them to find a job to help support their city lifestyle. Plenty of ant employers looking for work (aka scale insects and aphids) await in the trees we plant to line our neighborhood streets. These creatures depend on odorous house ants to protect them from ladybugs, tiny wasps, and lacewings, all aphid and scale predators. When odorous house ants show up, those predators split, enabling aphid and scale populations to soar. To pay for their security detail, aphids and scale insects provide odorous house ants with a sweet syrup called honeydew.

In the woods, odorous house ants compete with different species for places to set up camp. With acorn ants stuffing their homes into acorns, citrus ants pouring out from under tree bark, and acrobat ants peeking down from tree branches, odorous house ants make do wherever they can. But in the city, they can nest anywhere. Vacan-

Odorous house ants milk aphids like cattle for a sweet honeydew reward.

cies abound. From our garbage cans packed with odorous house ant–ready foods to the luxurious mulch we pile up around our homes to our kitchen floors, odorous house ants feast, raise babies, and have shindigs around us all the time. City odorous house ants can have many nests per colony with tiny superhighways of workers moving between them, distributing supplies from nest to nest. Some odorous house ant colonies can span a city block.

In the country, as conditions around their nests change, such as when a new, more dominant species comes to town or a big storm floods the area, odorous house ants move out. They generally move their nests every two weeks or so. This ability to pack up and move willy-nilly in the woods helps them cope with ever-shifting, human-made environments. Garbage day? Dumpster-living ants can saunter over to the grassy area. Dumping out those potted plants? Odorous house ants who had been living inside happily toddle over to the compost pile. Having many queens in the nest helps them split up without too many tearful goodbyes.

Roll Up the Welcome Mat

While I see odorous house ants in my kitchen as a happy surprise, I'm aware that not everybody (OK, probably not most people) shares my sentiment. It can be disconcerting to see eager sugarbears trundling across your Wheaties. After I conduct my wizardly identification, the response never seems to be: "What FUN!" It's almost always: "How can I get rid of them?"

Store shelves are packed with poisons designed to extinguish these ladies. However, knowing what we know now about odorous house ants, most of us can outsmart them. Be a detective. Stake them out. Follow them home to see how they are sneaking into your house. Then, eliminate the access point. We know that odorous house ants like to hang out in tree canopies and bushes, slurping up honeydew. Walk around your house and see if you have any bushes touching your walls or windows. Branches bridge the ants from their outdoor lifestyles to apartment living. Cut back those branches. Snoop

Odorous house ants carry larvae around the nest in their mandibles.

out other ways they enter the house. For example, they sometimes sneak in through cracks and crevices. Seal those with caulk.

We know they love to nest in mulch. People often dump piles of mulch around their homes. Switch that out for rocks, which odorous house ants don't like as much. Or try aromatic cedar mulch, which smells gross to odorous house ants, at least for a little while.

Look where they're crawling around inside, too. We know odorous house ants like sugar and all the delicious little treasures abundant in human garbage, so don't leave food out and tightly seal garbage cans. But even if you try to get rid of these sweethearts, pay attention as you do. Because the truth is, most of what might be known about these ants hasn't been uncovered. Most of their tiny empire's treasures lay undiscovered. So, while I can tell you as much as I've told you about sugar ants, I can't tell you much more. When someone calls to tell me about their sugar ants, most of what they have to report is not just grievance, it is science.

And so, when Ina says, "They keep stopping and talking to each other with their antennae," or Ariana reports, "I left my Coke open and they found it in less than 30 minutes!" these are things I write down, things you might want to write down too.

04 PAVEMENT ANT

SPECIES NAME: *Tetramorium* sp.E (the ant formerly known as *T. caespitum*)

SIZE: 2.5–4 mm (0.1–0.2 in.)

WHERE IT LIVES: Pavement ants most often nest under bricks or pavement, but they are also found in grassy areas near sidewalks and even in extreme environments, like salt marshes. They are heavily concentrated in Washington and Ohio, sprinkled throughout the Mid-Atlantic region, and occasionally found in the Southeast region.

WHAT IT EATS: Ultimate opportunists, pavement ants eat anything from dead insects to honeydew, a sugary food produced by sap-feeding insects like leaf hoppers. They also dine on pollen and food in your kitchen and garbage.

Wars happen across America every spring. Just as the trees begin to give us that first peek of color and the sun warms us enough to stretch our legs and venture outdoors for a look around, the animals begin stretching their legs, too.

Pavement ant battles are brutal. Here, four workers tear apart an ant from another nest.

Each spring, ants poke their antennae out of earthen holes, getting a feel for their new year on the beat. Workers of the pavement ant species (*Tetramorium* sp.E— although the pavement ant is common, scientists have yet to give this species a real name) push out of their nests with a mission: to establish their neighborhoods before ants from other nests nudge in and squeeze them out. These ladies are territorial, and they don't like any other ants walking on their turf. When they first emerge in spring, all the previous year's boundary lines have been wiped away with winter, and all bets are off. They redraw their property lines with warfare so gruesome it would make Attila the Hun blush.

Pavement ants are built for battle. At three-sixteenths of an inch, workers are about half as long as one of your shirt buttons is wide. They are dark reddish-black and have antennae that bulge out at the tips, making them look like they're waving little clubs from their foreheads. They have tough, armor-like skins called exoskeletons that can withstand the knocks of war. If a pavement ant were the size of a dog and you could get a good close-up look, you would see a beautiful landscape. Their faces and bodies are covered with hilly peaks, rivers of grooves and hairs, and they have two little mountains of spines poking out from their backs toward their rear ends.

Where neighborhoods overlap, huge numbers of workers from each side collide. They furiously drum one another's heads with

Each spring, pavement ant turf wars erupt on sidewalks.

their antennae; they rip one another apart with their mandibles. They'll separate an individual from the pack and close in around her, gnashing at her body with their jaws, grabbing her with their claws, turning her into ant dust. These ants mean business when it comes to setting boundaries. After the melee, the carnage is astounding. Thousands of ants litter sidewalks across the country, a jumbled dark line of body parts and pieces that blow around in the wind.

When they aren't out cruisin' for a bruisin', pavement ants move along slowly compared to other ant species, as though they don't have anything to do in this big old world but go for a walk in nature. They won't sting you, and they aren't easily spooked. Whereas some ants shoo away quickly, pavement ants usually continue to bumble along unbothered.

Pavement ants are not native to the United States, but they are one of the most common species around. They sailed over here in ships from Europe more than 100 years ago and flourish in the stone-slab environments of modern cities. They most often build their nests under bricks and in sidewalk crevices and will eat everything from sugary foods to dead insects to flower pollen to human garbage.

Sometimes, pavement ants act like miniature farmers. They collect seeds from plants and accidentally plant them by burying them in their nests. They also tend insects called plant hoppers like dairy farmers tend cows, "milking" them for honeydew, a sugary food the plant hoppers produce. If a plant hopper predator comes lurking around, pavement ants will pick the plant hoppers up in their mouths and carry them down to their nests, where they'll wait out the trouble.

Back to spring. The birds are practicing their songs, and you and I are hopping off the school bus, picking up lucky pennies, walking our dogs, or going to get coffee on our sidewalks that zig and zag from New York City down to Florida, across Tennessee, the Dakotas, and Wyoming, all the way to California. Each day, as we walk around in our world, the human world of sidewalks that point us to and from where we want to go, we are also walking over the world of the pavement ant, with devastating wars, property disputes, and peace times filled with farming and baby making. Their world is so similar to ours, so close to us, that we step over it every day without noticing how unusual these ants are.

05 FIELD ANT

SPECIES NAMES: *Formica pallidefulva, F. incerta*, and *F. subsericea*

SIZE: 5–10.2 mm (0.2–0.4 in.)

WHERE IT LIVES: *Formica pallidefulva* and *F. incerta* usually build mounds in the open, away from trees, while *F. subsericea* generally build their nests against trees, under rocks, or in logs. *F. pallidefulva* are found in South Dakota, along the Southeast and Mid-Atlantic regions, and in the Southwest. *F. incerta* are found along the Mid-Atlantic. *F. subsericea* are found along the northern Midwest region and the eastern half of the United States.

WHAT IT EATS: More buffet goers than picky eaters, field ants love sugar such as aphid honeydew, soft-bodied insects like caterpillars, and seed husks.

Formica ants, usually called field ants, are among the United States' largest and most common ants. Found spanning the states in all directions, three species make the most common list: *Formica pallidefulva* and *F. incerta*, both rusty-to-deep-red beauties, and *F. sub-*

sericea, black lovelies with stripes of sparse golden hairs across their rumps. Most field ants pass their days contentedly building their shallow, low-mound nests near rocks and trees, blissfully unaware of a dark underworld in their midst, a world of violence, slavery, mistaken identity, and poop shields.

About the size of one and a half pencil erasers, field ants' long, dexterous legs extend from their thoraxes, and their large black eyes rest right behind their always-moving elbowed antennae. You can reliably identify an ant as a field ant if it's a large ant, yellowish, reddish, black, or red with a brown or black rump. Many people confuse field ants with carpenter ants, neither of which can hurt you. If you'd like to tell if you have a field ant, gently nab the ant in question and check out its thorax, the middle section of the ant where all the legs attach. If its thorax consists of two lumps, you have a field ant. If it has one big hump, you're holding a carpenter ant. What a way to impress your friends!

Field ants have large eyes because they usually move around during the day and rely on sight more than some other ant species. They use those big eyes to help them see landmarks as they scurry to and from food. Like many ant species, field ants love tending aphids and scale insects for their sugary emissions, but they also help disperse plants by toting seeds around the forest, snacking on the husks and discarding the rest. They also enjoy wolfing down other insects whenever they get the chance.

Like other ant species, field ants tend aphids like cattle, milking them for sweet honeydew and occasionally gobbling a few like steaks.

Nice Outfit, Mr. Beetle

Field ants prefer to eat soft-bodied insects like caterpillars and beetle larvae, and this predatory tendency helps keep our trees happy. One of the northeastern United States' most dangerous forest pests is the gypsy moth. Thanks to their huge appetites, gypsy moth caterpillars have gobbled up more than 80 million acres of our northeastern forests in the past 40 years. When they scarf down all the leaves in the forest, trees die, causing millions of dollars' worth of damage. Fortunately, field ants love those plump little leaf munchers. They help reduce the damage and spread of gypsy moths by eating every caterpillar they can find.

Sometimes their partiality for pudgy little insects lands field ants in unusual situations. Many baby beetles (often called grubs) fit the mold for a perfect field ant meal. Slow, soft, and chubby, beetle grubs don't stand a chance when hungry field ants stumble across them while foraging. To ward off potential beetle slayers, many beetle species, like tortoise beetles, rely on an inventive solution: poop shields.

Here's how it works: Some plants in our forests and across our cit-

ies have certain "stop eating me!" chemicals in their leaves, called deterrents. When most insects bite into a leaf and smell the deterrents, they get as far away as possible. Not our resourceful beetle grubs. They eat as much of these stinky leaves as they can, pooping stinky leaf poop all over the place. Then they gather up the poop and stick it on their bodies, making a force field of stink that follows them wherever they go. Field ants catching a whiff of these otherwise tasty tidbits run in the opposite direction of our little Pigpens. If you feed these baby beetles nonstinky plants, they still make a poop force field, but because it contains no deterrents, field ants will ignore the BM blanket and eat them right on up.

Slaving Away

It may seem like all fun and games for field ants, frolicking across our forests, lawns, and traffic medians, grocery shopping and building their houses. But field ants have a wicked foe prowling those same forests, lawns, and traffic medians, combing the grass for field ant nests: Amazon ants. Amazon ants look a lot like field ants: same size, similar color, same big eyes, similar camel-humpy back. Amazon ants and field ants look so similar they could pass as the same species, almost: Amazon ants have dagger-sharp, sickle-shaped jaws. Their jaws are so pointy they can't take care of tender babies—any attempt at carrying or feeding could result in a fatal stab wound to their young.

So Amazon ants came up with a solution: They raid field ant nests, frighten adults into submission with poofs of chemicals, snatch up hearty pupae in those jaws, and scurry back to their nests. Now, we remember from the ant's life cycle that baby ants take a lot of food, but once those ants pupate, they don't eat at all. They just sit there helpless in their nests and wait to turn into adults. By stealing pupae, Amazon ants basically snatch up soon-to-be adult workers that require no maintenance in the meantime.

Amazon ants carry slaves, unlucky field ant pupae.

Once in the raiders' nest, the field ant pupae start to pick up the smells in the nest. Ants tell one another apart by smell. If a field ant starts to smell like an Amazon ant, she'll start to think of herself as an Amazon ant. When she emerges as an adult, she will do the tasks to help the colony that she would have done in her real mother's nest: gathering food, building the nest, raising babies, taking care of the queen. She usually will have no idea that she's a slave, helping her enemies to grow so they can raid more field ant nests.

Each summer, poor field ants are enslaved up and down the United States, from the forest near my North Carolina house to the parks of busy Long Island, New York. But you and I can still spot those lucky enough to have escaped the dagger-jaws of the Amazon ants. They run along our tree trunks and across our sidewalks, planting seeds, snagging bugs, turning soil. We can look for their double humps and drop them a snack like a piece of a cookie or some juice from our juice boxes and see if they eat it. We can give them advice, telling them to stay away from poison poop and to keep their big eyes peeled for slave makers. They can give us advice, too. They can tell us never to underestimate the power of small things, to be mindful of the good they can do. They can tell us that every animal has a complicated story, a life of adventure and trials that unfolds whether or not we humans pay attention. But you and I can pay attention. Field ants are happy to share their story with us.

06 LASIUS ANT

SPECIES NAMES: *Lasius neoniger, L. alienus, and L. niger* (a common species, though not yet documented in Chicago)

AKA: small black ant, cornfield ant

SIZE: 3.0–4.2 mm (0.12–0.17 in.)

WHERE IT LIVES: *Lasius* ants prefer open spaces and set up their anthills in grassy areas like golf courses and traffic medians. They sometimes nest under stones or in logs. *Lasius alienus* are heavily distributed along the northern half of the West Coast, are found in the Southwest, and are sprinkled along the East Coast; *L. neoniger* are found throughout the United States; and *L. niger* are found along the West Coast and are heavily distributed in Utah and New Mexico.

WHAT IT EATS: Aphid experts, *Lasius* ants tend aphids like cattle, milking them for honeydew and sometimes killing them for a big aphid steak.

Before I met my husband, an electrical engineer, the only thing I knew for sure that engineers did was drive trains. Sure, I knew tons of people who said they were engineers of one kind or another, but

once I found out they weren't wearing cool hats guiding thousands of tons of steel hurtling at lightning speeds down railroad tracks, my brain would glaze over and I'd lose interest in their profession. "You mean you have no giant horn to blare at passersby as you cross over city streets and along the countryside? BOR-ing!" I used to think.

Then I met the man who would become my husband. As he talked about his job, I learned that many different kinds of engineers work to keep our world safe and running smoothly. Civil engineers, for example, prevent tragedy by designing safe roads, buildings, and bridges. Electrical engineers plan the circulatory system of wires and currents coursing through our cities. Aerospace engineers launch us into the sky and even into outer space. Many more types of engineers tool away behind the scenes, perfecting our lives without our noticing.

In nature, some ant species work as engineers, shaping and refining the environment, building connections, repairing bonds. *Lasius* ants are one group renowned for being superior soil engineers. A little larger than a sesame seed, *Lasius* (LAY-see-us) ants look more like regular ol' ants than the movers and the shakers of nature. They resemble odorous house ants with slightly larger (and fuzzier) behinds and are generally darkish brown to deep brownish black in color. While odorous house ants move like troopers marching in line to and from their favorite foods, *Lasius* ants move more like my favorite Aunt Ann: deftly, but with a bit of a waddle, probably owing a bit to those fuller fannies.

Lasius nests often look like mini-volcanos erupting from the grass.

Three *Lasius* species make the most common list: the small black ant (*Lasius niger*, not yet documented in Chicago), the cornfield ant (*L. neoniger*), and *L. alienus* (no common name yet). All *Lasius* species have only one queen per colony and prefer to nest in soil in open areas or under logs and stones. Oftentimes, their nests look like little volcanoes popping up across grassy areas, and, since one nest can have many entrances, they can look like an erupting mountain range spreading out across the landscape.

Lasius ants are often the most abundant ant species on golf courses, happily setting up shop in the expansive open habitat that seems tailor-made just for them. These ants' abodes get them into trouble with golfers, who would prefer their putting greens smooth and free of ant-made speed bumps. But researchers studying *Lasius* ants have shown that when exterminators try to smooth the greens by poisoning *Lasius* ants, more destructive golf course pests like Japanese beetle larvae and cutworms thrive.

In addition to the rolling green, *Lasius* ants love golf courses because they love sugar. Yes, these pristine courses may be free of half-eaten Snickers bars, but *Lasius* ants have their eyes on a different type of sugary treat: chubby, sugar-making insects called root aphids who slurp away at underground grass roots. *Lasius* ants show off their extra-engineering talents as farmers by protecting these root aphids from predators with little mouth-built sheds, all the while milking them for their sugar.

Lasius ants are "aphid experts." They can climb up a tree and tell which aphids like ants and which don't just by sniffing with their antennae. No dummies when it comes to sugar, they prefer the sweetest sugar around. They'll walk around trees sniffing aphid butts until they find the species that produces the sweetest sugar. Whoever wins the sweetest award gets protected and tended by the ants. They're so good at kicking out predators that a ladybug will avoid laying eggs in any area where she even catches a whiff of *Lasius* ants.

Just as some farmers need to kill part of their cow herd to eat meat, *Lasius* ants need to kill some of their aphid herd for protein.

Watch ant nests long enough, and you're sure to see a resident emerge, like this *Lasius* ant.

How can you choose your best beef from all of your best-producing Bessies? If you're a *Lasius* ant, you follow your nose. *Lasius* ants eat aphids from their herd that they or their sisters haven't tended. They can tell who's been milked by sniffing the aphids to see if they've been touched by a member of the colony.

But enough about sugar. Back to the engineering. *Lasius* ants engineer soil. To understand how they help develop our dirt, we first need to understand a little bit more about good ol' terra firma, which props us up right this second, whether we think about it or not. Even though dirt might seem as dead as dead gets, healthy soil actually lives and breathes just like we do. As dead plants and animals decompose, they release nutrients and gasses. Microorganisms then scoot around gobbling up some nutrients and turning them into other nutrients. Animals like earthworms push the dirt about, letting air flow through, speeding up biological processes that increase life-promoting properties within the soil. When nobody contributes to these processes, plants and animals can't survive because the soil gets compacted and hard, with nutrients concentrated in some places and scarce in others.

Lasius ants are among the first species to set up their tents in disturbed areas, and from the moment they move in, they start turning the soil and making the ground ready for life. They burrow their tunnels into the soil, tripling soil respiration, the soil's ability to "breathe," as compared to non-*Lasius* areas. They do such a good job at aeration that more insects and spiders move into those *Lasius*-affected, loosely compacted, fresh-air areas, which has a cascade of positive effects on the ecosystem around them. They carry dead insects and sweet treats from one place to another, spreading out some soil nutrients and concentrating others, which increases all kinds of chemical and biological processes. They even likely help fix nitrogen, an important chemical for healthy plant growth. Just like the engineers who build our world and help it flow, *Lasius* ants churn and build wherever they live.

When was the last time you rode over a bridge or walked into a building and thought, "Gosh, isn't it nice? I'm not afraid this building/bridge will collapse on me"? Probably never. We don't think about it because we don't have to. Thanks to our engineers, things work.

The same goes for soil in our natural areas. We seem to pay attention to our soil only when it stops working well—when plants die, animals move out, and the soil actually becomes stone dead. Our *Lasius* engineers, like my husband and the engineers that reinforce and shape our human-made world, don't seem to mind that we don't notice. They don't need us to shower them with thanks, but they do need us to give them the space and resources they need to do their best work. Keep your eyes peeled for volcanoes on the putting green or in your city's medians. See if you can spot a sesame seed waddling out. Give it a salute as it passes by, conducting its business of building and bettering our world.

07 TRAP-JAW ANT

SPECIES NAME: *Strumigenys* spp.

SIZE: <2.54 mm (<0.1 in.)

WHERE IT LIVES: Tiny trap-jaw ants live in rotting logs, under stones, and in the soil—anywhere near their favorite meal: collembolans. *Strumigenys rostrata* are primarily found throughout the Southeast and Mid-Atlantic regions.

WHAT IT EATS: Stealthy-but-blind hunters, trap-jaw ants stalk between grains of dirt for little soil dwellers like collembolans, mites, and termites. But their favorite meal is collembolans.

It's dark. You're surrounded by stones and boulders. Huge green vines thrust and spiral outward from the earth. Woody mountains knot the terrain all around you. Springtails—large, soft creatures—peacefully push the rocks with their antennated foreheads and ten-

Trap-jaw ants are so small that, to them, moss fronds are a forest.

der, fingery legs, foraging like cattle for decaying plants under the rubble. Somewhere behind the boulders, a long-toothed predator lurks.

In the shadows you can just make out its dagger-jaws, gaping and ready for its next meal. You see its narrow, skull-like cranium pitted and dotted with sparse hairs, its slender, knobby, stealthy killer's body, grooved and hairy, tipped at the end with its deadly stinger.

You watch in the darkness as it tip-tip-taps its own antennae one way and then the next, mandibles locked open, using scent to stalk the soft-bodied creatures around. The predator moves cautiously, deliberately, getting closer, closer still to its unsuspecting meal. Its antennae lightly sweep its prey; its jaws sneakily surround the victim's body. When the prey moves, even slightly, against those waiting jaws: CRACK! In less than two-thousandths of one second, the hunter's spiked mandibles snap shut around its quarry, impaling the soft flesh. The poor animal tries weakly to struggle, but the predator lifts it high in the air and stabs it with a venomous sting, paralyzing it for the journey home to the predator's nest, where its sisters will disassemble the meal for a family feast.

This resident assassin lurks in the darkness all around you, but you don't need to be afraid. The *Strumigenys* ant, our sickle-toothed hunter, measures less than one-tenth of an inch—nearly as tiny as the grains of dirt between which she navigates. From our human-size point of view, she is almost too small to see without a magni-

Tiny springtails are a trap-jaw's favorite dish.

fying glass or microscope, so small you could pack half a colony of her kind into a buttonhole. And while she may strike terror in the tiny hearts of soil dwellers, to me she looks like an unearthed jewel, something rare and adorned and stunning.

Strumigenys, the tiniest type of trap-jaw ant, whose name literally means "tumor jaw" for its large, powerful mandibles, is common in moist soil, rotting logs, and under rocks across the United States. However, many people don't realize ants like these live underfoot. Because of their diminutive size and their soil-dwelling lifestyle, these ladies can be hard to spot as they cruise the underworld for springtails. Two species make the most common list, *S. talpa* and *S. pulchella*, but many more *Strumigenys* species stalk the soil around us. Although some *Strumigenys* will settle for juicy mites or other soft-bodied soil dwellers, most prefer a springtail supper.

Springtails, also called collembolans, are like the Holstein cows of the undergrowth, enormously abundant and fantastically nutritious. Not quite insects, springtails look like insects that haven't quite been fully formed yet. Some are round and fat and almost could be mistaken for miniature aphids, while others have longer bodies with peaceful black eyes and long, bouncy antennae. All springtails, no matter how slender or plump or bald or hairy, have long tails, called furculae, which they tuck under their bodies. When a springtail senses danger, it will pop its furcula down like a spring, striking it against the ground, propelling its body high in the air—sometimes to a distance almost 100 times its own body length. Because of this emergency escape route, *Strumigenys* need to be super stealthy when on the hunt for springtails, and these ants adopt a distinctive, creeping, sneaking gait that makes them seem like they're always up to something.

Nearly blind, *Strumigenys* ants rely on their sense of smell to locate and capture springtails. Because springtails can't see too well either, they often accidentally crawl right over unsuspecting *Strumigenys* backs without the ants noticing they've missed a good meal. The ants also consume mites, tiny termites, and other minute creatures crawling through the dirt. When such a creature happens by and grazes the trigger hairs between a *Strumigenys*'s jaws, the jaws snap shut. Sometimes, a *Strumigenys* ant will sneak her jaws around a potential meal, only to find the meal doesn't bump into those trigger hairs. If this happens, she'll wait for a few minutes in case the bug wants to wake up. If it continues to sit still, she will tap its back with her antennae to try to wake it up to an unpleasant surprise.

Tiny trap-jaws live in correspondingly tiny nests excavated out of rotted wood or damp soil, sometimes under stones or leaves. They usually have between 25 and 100 workers crawling through their nests. We don't often see them if we're looking for ants on the pavement or our front porches, but they're there, wandering through the soil and leaf litter. We can find them by stealing a bit of leaves and

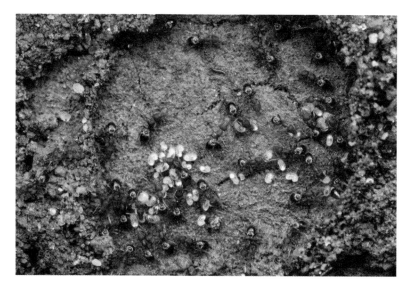

08 PONERA ANT

SPECIES NAME: *Ponera pennsylvanica*

SIZE: 3–3.75 mm (0.12–0.15 in.)

WHERE IT LIVES: Hidden treasures, *Ponera* ants sneak just below the soil's surface beneath leaf litter. They are distributed throughout the eastern half of the United States and have also been found in the Midwest and Southwest regions.

WHAT IT EATS: *Ponera* ants dine on tiny soil invertebrates.

Head outside in late spring through early fall with a flat-bottomed pan and a sieve from your kitchen. Find someplace that doesn't get too dry, maybe a shady spot with some good rotting logs. Scoop up a little leaf litter and topsoil in your sieve. Then shake it over your

pan like you're mining for gold and see what falls out. This is how ant researchers mine for their own gold, those priceless little creatures that, because of their minute size or their habit of creeping just below the soil's surface, are easily missed by standard ant-gawking techniques. For ant researchers, this "ant gold" includes thief ants, those pint-sized burglars, or the long-jawed tiny terrors, *Strumigenys* ants.

Sometimes, if you're lucky, a clunky brown-black nugget will plop down in the pan. No fatty by any standard, this diminutive creature is more of a "big boned" ant. With little sign of the waspy waist most delicate ants boast, this mini ingot looks like one long anty tube. Watch her explore your pan. With her chunky middle and thick, clubbed antennae, she moves like one of those toy trains, hooked together engine to caboose, that topple over rather than turn on curves, almost like she's just figuring out how to propel herself forward. You hit the big time with this one. You found a *Ponera pennsylvanica*, or *Ponera* ant.

Like many of the ubiquitous-but-less-studied curiosities making their living under our toes, *Ponera* ants have no common name. Also like their less-studied cousins, most of the better-studied ants dwarf them in size, with *Ponera* ants often measuring at almost a tenth of an inch, about the size of two poppy seeds. They prefer moist soil and dig their tiny nests in soft, rotting logs, old acorns, and decaying leaf piles in forests and semi-urban areas across the eastern and central United States, from Canada to Mexico. Excellent predators, *Ponera* use their long, venomous stingers to stun and kill tiny soil-dwelling arthropods whenever they want a good meal. Though we don't study them often, *Ponera* ants can teach us many things about ant history. They're walking, breathing examples of primitive ants.

Now to ant researchers, the word *primitive* doesn't mean these ants are cavemen equivalents, with rudimentary tools and comparatively undeveloped brains. It only means they share many characteristics with their earlier ant ancestors and can help us to see how ants must have looked and behaved a long time ago, before we were around to give them a second thought.

A little over 90 million years ago, right around the time plants began to flower for the first time, wingless, wasp-like creatures found their niche in the dinosaur-stomped earth. They prowled treetops and under soil for a good meal. They were just beginning to figure out a social lifestyle. Queens as we know them hadn't yet made their big ant debut, so ant workers could still lay eggs.

Ponera ants have much in common with those early ants who

Ponera ants carry pupae and do other nest chores.

were just settling down for family life. For starters, they have very small colonies. Compared to other ant species with newer survival strategies like Argentine ants or red imported fire ants, which can have colonies with thousands, millions, trillions of workers, *Ponera* ants max out at just a few workers, with rarely more than 50 individuals in a whole nest. They grow their colonies at a slow, steady pace with just a few workers added each year, sometimes only enough to replace those lost to hungry ant munchers or people's clompy feet.

Unlike for these more modern-seeming ant species with their big, fat queens—whose colonies have plenty of room inside for baby making—it can be hard to tell a *Ponera* queen apart from her workers. And, like those earlier ants, if something happens to a *Ponera* colony's queen, *Ponera* workers may be able to start to lay eggs.

Chemicals are ants' true eyes and ears. You and I hear the garbage truck barreling down the street; we listen as our friends whisper gossip on the telephone; we watch the bright flash of birds as they flit by. But ants can be very hard of hearing (they have no ears, though they can sense vibrations), and most ants can't see all that well. Their colors and language come from invisible chemicals that ooze, puff, spread, or squirt from more than 10 glands on their bodies, and they receive signals from a similar number of receptors, tiny taste buds here and there on their bodies.

Some of today's ants, like *Formica* ants, have mastered their chemical language. They can recruit thousands of workers to a drop of orange juice in minutes; they can call troops en masse to attack an enemy; they can mark their home and the territory around it; they can "talk" to each other about what types of snacks they found, how abundant those snacks are, and where to find them. These primitive-seeming ants, on the other hand, speak a more direct-function language.

Ponera ants still have chemicals that signal "Alarm! Alarm!" Rather than telling their sisters to round up and attack, *Ponera* ants' alarm means "RUN!" Any disturbance can cause them to haul tail and hide

as quickly as possible. Hiding works better if there aren't enough colony members to mount a hearty attack. With *Ponera* ants' small colony size, an attack would probably result in complete devastation, while running away and hiding could save the nest.

You may be used to watching ants follow their chemical trails to food all over town, but *Ponera* ants don't have these elaborate chemical highways. Instead, they either hunt alone or follow their sisters directly to the food, with the follower tapping on her sister's rear end with her antennae to make sure she stays in contact. She and her ancestors have been tap-tap-tapping a long time. Her signals may be simple but they have been enough to allow the species to persist for tens of millions of years and still be one of the most common ants.

09 WINNOW ANT

SPECIES NAME: *Aphaenogaster rudis*

SIZE: 3.81 mm (0.15 in.)

WHERE IT LIVES: Winnow ants prefer to nest in rotting wood but will nest anywhere from soil in open areas to human garbage. They are distributed primarily throughout the eastern half of the United States but have also been found in Wyoming and Colorado.

WHAT IT EATS: Winnow ants eat the tasty outer coatings of seeds, and other insects like termites. They also like sugary foods.

Aphaenogaster rudis sounds more like an unsavory medical condition than one of the coolest ant species in North America. The name doesn't roll off the tongue like "sugar ant," "carpenter ant," or "pavement ant." So, for the purposes of familiarizing you with one of the

best residents on your block, we'll give *Aphaenogaster rudis* a nick-name: the winnow ant.

Winnow ants are among the most elegant-looking ants around the forest and in your backyard. With their long legs and slender reddish-brown bodies (leading to the nickname they share with their cousins: thread-waisted ants), they pick their paths delicate-ly across the ground like rusty ballerinas. Each medium-to-large worker measuring at about 0.15 inch can just cover the date on a quarter. Although they prefer to nest in decomposing stumps and logs, winnow ants can make the best out of any situation, building their homes in open soil, beneath rocks, and even in human garbage.

Cracking open forest logs in the spring often reveals a hidden world of winnow ants.

With one queen and up to 2,000 workers, a winnow colony could easily pack a stadium for an ant rock concert.

Beyond their refined appearance and wide-ranging nesting habits, winnow ants have two qualities that set them apart from the rest of the ants: the helping hand they give forest plants and their ability to use tools.

First, let me tell you about their agricultural talents and the reason we call them winnow ants. Winnow ants have a special relationship with forest plants. We all know that many plants make seeds. Some plants produce seeds with a special coating called an elaiosome that's a lot like the hard candy coating on the outside of an M&M. The elaiosome has a special blend of flavors that is irresistible to winnow ants.

As they pick across the forest floor in search of food, winnow ants often stumble across these seeds. When winnow ants get a whiff of that elaiosome, they can't help themselves: They have to pick up the seed and carry it back to their nests. Once in the nest, winnow ants feed the outer coating of the seed to their young.

Winnow ants munch on a seed's tasty outer coating before planting the seed.

Unlike most of us, who prefer the chocolaty center of M&Ms, winnow ants eat only the elaiosome and leave the seed inside alone. When wheat farmers remove the chaff from wheat seeds, it's called winnowing. Likewise, winnow ants remove the chaff from forest seeds. Once the elaiosome is gone, the ants don't need the seed anymore, so they take it back out of their nest and deposit it on the forest floor or in fertile underground middens. There, the seed, no worse for wear, is free to sprout and grow into a happy forest herb. Having their elaiosome nibbled away for hungry ant babies does not hurt the seeds; in fact, it helps them. When ants take these seeds back to their nests, they in effect protect them from animals that eat the whole seed. Later, winnow ants "plant" the seeds (by discarding them outside the nest or in middens) far away from the seeds' parents. As a result, the newly planted seeds don't crowd their parents as they grow.

Seed planting is a successful business for winnow ants and the seeds they plant. Almost two-thirds of all herb seeds produced in the forest, such as from wild ginger and trillium, are picked up by winnow ants. Also, when winnow ants are removed from forests, some wildflower abundance drops by 50 percent. Seed planting also helps the ants. When winnow ants eat that candy coating elaiosome, they get all the nutrients they need to make more babies.

Farming isn't the winnow ant's only talent. Like other animals, from woodpecker finches to chimpanzees to humans, winnow ants use tools to gather food. When a winnow ant happens upon liquids too goopy to carry back to her nest, she goes out in the forest and collects bits of soil, leaves, and sticks. She takes these bits back to the newfound food and drops them right on top of it. These leaves and sticks become little plates for winnow ants. Workers bring the plates back to the

10 LITTLE BLACK ANT

SPECIES NAME: *Monomorium minimum*

SIZE: 2 mm (0.08 in.)

WHERE IT LIVES: Little black ants can make their nests outdoors, in forests, or right in our backyards, often under rocks and tree bark. They are heavily distributed in the Southwest, Southeast, and eastern Midwest regions. They have also been spotted in the continental United States as far west as California, as far north as North Dakota, and as far northeast as New York's southern tip.

WHAT IT EATS: Little black ants eat a sugary liquid called honeydew, made by small insects called aphids and scales. They also eat dead insects, spiders, and your trash.

When I was little, we had a nest of what I later learned were little black ants under a cherry tree in our yard. In the thick of summer, tired of digging foxholes all over the yard, my brother Will and I

would follow them. We'd grip the trunk with our monkey toes and climb to the outside branches. We'd lean out as far as we could to see what they were up to. So long as we stayed out of their way, they never seemed to mind. They went about their business beneath the leaves and around the branches as we went about ours.

Little black ants are among our cutest "most common" ants. As their common and scientific names suggest, little black ants are much smaller than many of the other ants you see hanging around your house and yard. Their glossy sheen adds a touch of determination to their comings and goings. It's as if they take themselves too seriously, little polished wingtips toddling to and from their important ant business.

In our ant stalkings, Will and I quickly learned that, while they might be fun to watch, little black ants could be little jerk ants if we interrupted their work. Although small, these ants would jab their tiny stingers in our thighs and arms if we accidentally blocked their trails. It doesn't hurt much, but it's enough of a reminder to keep moving.

Little black ants bully other ants over food resources. Their colonies can number more than 2,000 workers, and when they get upset, they recruit their sisters in high numbers. When little black ant workers combine forces, their tiny stingers can pack a powerful punch to other ants. They put their best combat skills on display when they protect one of their favorite foods: the sweet nectar produced by sap-sucking insects such as aphids.

With mouths shaped like drinking straws, aphids live on plant leaves. They stab these straws into the leaves and suck out the juice like it's a big milkshake. They then turn that juice into honeydew that they excrete from their rear ends in droplets they hold high in the air, waiting for ants to come and get it.

To reach their favorite syrupy snack, little black ants travel in long lines up tree trunks and plant stems. They make the line by laying down scented pheromone trails. Even after their sisters are gone, the trail remains, a scented road to good food that they follow by waving their antennae back and forth over the path.

In addition to making delicious honeydew, aphids are tasty snacks for other insects like lacewings and ladybugs. But scrappy little black ants kick out other would-be diners from their honeydew buffets, even if those diners dwarf our tiny, shiny brawlers. Despite their Lilliputian size, little black ants' stingers and chompy mandibles can inflict more damage than other, larger ant species. With little black ants around to keep predators away, aphid numbers increase up to 10 times their normal abundance.

To little black ants, grains of sand are boulders.

Although up to the challenge in groups, when little black ants get caught alone, they have other options. Suppose a fire ant finds a little black ant hanging out on a root and decides to pick a fight. Instead of fighting back, our little black ant will "flag" her gaster (her abdomen), wagging it around in the air as if to say, "You'd better stay away from me! I mean business!" While she wags, she will release noxious toxins, hoping to repel the contender before she has to fight.

If booty-shaking fails, our little black ant will curl up and act dead, playing possum in the hope that the fire ant will think herself victorious and just go away. Sometimes, little black ants combine their individual possum-playing and group brawling behaviors to persist in areas with more dominant ant species. These little ladies can even push out fire ants trying to move into their neighborhood. They snack on fire ant babies as a reward for triumphant battle.

When Will and I watched the little black ants twine around our cherry tree on those hot summer days, fire ants had not yet made their march into North Carolina. Little black ants were the only game in town on that side of our house, with carpenter ants and field ants galloping through the front yard and high noon ants staking their claim to the hard-packed dirt and centipede grass in the backyard. Will's a grown-up lawyer now; his monkey toes spend the day in dressy shoes. The cherry tree was cut down 20 years ago, its last fat fruits still clung to the branches all piled up on the curb. But little black ants are still the same. When I find them greeting me on the walkways of campus or snaking across my porch, their shiny heads determinedly pushing forward, I fill up with the pleasure of seeing old friends. When we understand these elements of nature, get to know them by name and habit, we will always be surrounded by friends.

11 THIEF ANT

SPECIES NAME: *Solenopsis molesta*

SIZE: 1.5–1.8 mm (0.06–0.07 in.)

WHERE IT LIVES: Thief ants nest underground in forests and open, grassy areas. They also like to nest in human structures. They particularly like nesting near other ant species' nests. Thief ants are heavily distributed in the northwest tip of the United States, in California, in the Southwest region and sprinkled throughout the eastern half of the United States.

WHAT IT EATS: Tiny ants with big appetites, thief ants prefer protein, such as dead insects.

Back in the days of the Wild West, Jesse James and his outlaw gang were some pretty crafty dudes. They robbed everything from stagecoaches and trains to banks and homes. His bandit bunch crept into

towns and would hightail it out ahead of angry lawmen and WANTED posters bearing James clan faces. Imagine if the Jesse James family moved in right next door to your house! Many ant species across the United States face this predicament every day when thief ants come to town. Thief ants are the Jesse James gang of the ant world, and these bite-size burglars pickpocket and plunder anything they can get their little mandibles around, living lives of artifice that would make Mr. James sit up and take some notes.

Even though he was a robber and a murderer, Jesse James won the public's hearts, in part because he was easy on the eyes. Thief ants are no different. Whenever I stumble upon a thief ant nest or happen to lift a dead insect and find a bevy of thief ants, mid-snack, I always stifle a squeal. Thief ants are unbelievably, ridiculously cute.

Their size might play a big factor in their cuteness. At one-sixteenth of an inch, a thief ant worker could wander comfortably around in a lower case *o* on this page. Most often a golden yellow color, thief ant workers vary along the color spectrum all the way to amber. They have stingers, but they are too tiny to cause you any pain. They look like they wander around really slowly, but actually they're just super small. If you had a microscope, you could see that each antenna has a bulb on its end, and they bonk about as they feel their way to and from food. Much of that food, remember, is stolen, either from other ants or from you and me.

Thief ants get their name from their habit of setting up camp next to other ant species' nests. They love protein and stuff their bellies on dead insects, people food, and insect eggs. When the other ants bring home thief ants' favorite foods, those crafty little burglars sneak that food right on over to their own houses and feast. They've also been known to smuggle out other ants' babies, tasty snacks for greedy thief ants. When other species' colonies are weak or dying, thief ants aren't as sneaky. They run through the nests' halls like children running down the aisles of a Toys"R"Us on a shopping spree, eating their fill of dead and dying ants.

Their crimes and misdemeanors don't stop with the insect world: A thief ant will rob you blind if you don't watch out. Thief ants are opportunists, and they recognize that your kitchen is a wonderful opportunity for the biggest heist of their lives. Because they are so small, many people have a hard time figuring out how to keep them out of their pantries. The best way to keep thief ants out is to figure out how they're getting in. Once you do that, block their entranceway by plugging holes with some caulk or weather stripping and tell those thief ants there's a new sheriff in town.

Some people think Jesse James was like a modern-day Robin Hood and that many of his crimes were to benefit others. I don't know what Jesse did with all of his loot, but many of the thief ants' crimes against other insects surely do help us out a lot. For example, when they're not stealing from other ants, they love to eat lawn pests like cutworms and scarab beetle eggs, and they provide effective control against these lawn and golf course pests.

Even though they're miniscule, they're pretty good at bullying one of the southern and western United States' biggest ant bullies: the red imported fire ant. Thief ants are almost three times smaller than the smallest fire ant. Like the James gang, they rely on their cunning and strength in numbers to beat up and eat any upstart fire ant colony making camp in their territory. In fact, fire ants can't establish nests in areas where thief ants roam.

Being tiny has its advantages. Because thief ants nest underground and out of sight, they are one of the few ant species who can weather the havoc wreaked by other nasty invaders like Argentine ants and yellow crazy ants. When other ant species get kicked out of town, thief ants hold their ground.

Jesse James's shoot-'em-ups and looting sprees came to an abrupt end when he met the wrong end of Robert Ford's pistol. Fortunately, thief ants survive even the toughest ant assassins. They beat up and outwit other invading ant species. Unlike Jesse, who caused trouble with the law wherever he went, thief ants contribute to our natural world. They help keep other pieces of nature in check by eating dead insects and aerating the soil with their underground nests. You could even say they are lawmen in their own right, nibbling away at the pests crawling around your lawn. They're tiny but tough, and they're outside your door right now. Despite their name, thief ants live mostly on the good side of the law.

Thief ants tend root aphids underground.

12 WINTER ANT

SPECIES NAME: *Prenolepis imparis*

AKA: false honey ant

SIZE: 3–4 mm (0.12–0.2 in.)

WHERE IT LIVES: Winter ants nest deep in the soil near tree bases or in open ground, like lawns. They are heavily concentrated along the West Coast, sprinkled throughout Florida, and found in the Midwest, Northeast, and Mid-Atlantic regions.

WHAT IT EATS: While winter ants won't pass up an opportunity for a sugary snack, these ladies prefer protein-packed food, noshing on other insects not lucky enough to endure winter's chill.

Remember in *Alice in Wonderland* when Alice followed the white rabbit down its bunny hole? The hole was ordinary enough at first, but once Alice climbed in, she fell down and down until she came to a completely different world. Holes like that rabbit's pepper the

ground across the United States. If we were as small as ants, we could tumble repeatedly down into other worlds. Winter ants are the white rabbits of ants. Plunging down their holes gives us a peek into their truly extraordinary lives.

Unless you follow a winter ant home, its nest's entrance can be hard to find. About the size of a buttonhole, winter ant nests aren't a lot to look at on the outside. Inside, however, deep mazes of tunnels connect chambers all the way to the bottom. The nests can extend almost 12 feet deep into the soil. That would be the human equivalent of a class of second-graders digging a hole more than 1.14 miles down, deep enough that 150 school busses could be stacked end-on-end before reaching the surface.

All that depth serves a purpose. While most ants are active in the spring and summer, winter ants prefer the fall, winter, and spring. Soil temperature does not vary as wildly as the temperature above ground, so when winter's chill plummets to 33°F, the winter ant's nest is kept insulated by the earth and remains at a balmy 64–68°F. This heat is important because between 40–50°F, most insects develop a serious case of brain freeze, going into what bug people call a "chill coma," where their muscles stop working so they can't move. Underground, winter ants beat the ice. Above ground, they dig short "warming tunnels" scattered around their nest. When they start to get cold walking around outside, they run down into the tunnels and warm up.

Staying Out of Trouble

My mother always told me the best way to stay out of trouble is to avoid it. Winter ants are masters at avoiding trouble because they move about when trouble is fast asleep. From March to November, when most ant species scramble around gathering food and fighting one another for space, winter ants seal themselves tightly in their nests. When November rolls around and other ant species tuck themselves in for their winter naps, winter ants unseal their nests and begin exploring the world. Because they are active when other ants sleep, they often miss

the dangerous tides of invasive ants that can wipe out many other ant species. In this way, they can persist in areas inhabited by other inhospitable ants. If they do happen to meet an adversary, they spray a toxic chemical from their rumps that scares off or even kills the would-be contender.

How to Spot Them

At the beginning of fall (or winter, in warmer climates farther south), winter ants are hard to identify. Shiny reddish-brown with lighter yellow legs, they look like your everyday, run-of-the-mill ant. Early in the season, workers are about 0.1 inch long, just long enough to span the letter *t* on this page. But as the season progresses and winter ants stock up on food, they become easier to identify.

Sometimes called "false honeypot ants" for their large rear ends, winter ants use that extra junk in their trunks to survive summer underground.

To say a winter ant has a big behind mid- and late season is an understatement. When workers eat their favorite protein foods like insects and a sugary substance produced by other insects called honeydew, they stockpile the calories in special fat cells in their bums. These fat cells can grow to be of tremendous size. Because they waddle around with swollen rumps at the end of the season, some people call winter ants "false honeypot ants."

To understand why they pack on the pounds, let's poke our heads down into one of their rabbit holes. Winter ant colonies survive underground all summer on their rotund sisters' fat. Their ample behinds are the world's best refrigerators. Because fat cells are part of living tissue, as long as the worker is alive, the fat won't rot, like dead insects stored in the nest would. And because the fat is already concentrated and high in calories, workers don't have to process it like they would other foods. Winter ants store enough fat in their portly posteriors to feed each other and all the babies that emerge as adults the following fall. When workers unseal their nests in the fall, they emerge as Skinny Minnies again. In colder climates, they may become active at night in the summer, getting a jump on gathering food before the weather's too chilly for even their fancy footwork.

Home Deep Home

Let's travel a little deeper down the rabbit hole. While its worker inhabitants live a couple of years at most, winter ant nests can exist for more than ten years. The older the nest, the deeper it is. If we were winter ants crawling down into our home, we would enter through a short hallway leading to the first room. Other than the pinhole of light shining through the entrance, the whole house would be completely dark. To get from room to room, we'd have to smell our way with our antenna. Our rooms would have domed ceilings, tall enough for a couple of us to stand on top of one another. Because we'd have clingy feet, we could even walk on the ceiling!

We might have a few hundred sisters—sometimes up to 10,000—living with us, so every now and again, we'd bump into one of our sisters and give her a friendly tap with our antennae. If she seemed hungry, we might spit up a bit of food for her to eat. If she seemed dirty, we'd help clean her with our mouths and antennae.

It might take us a long time to get all the way to the bottom of the nest. Remember, winter ant nests are at least the human equivalent of a mile deep. Our older sisters would live in the upstairs rooms, and our younger sisters would live with our mothers deep down. Our queen mothers would wander around the bottom of our nest in the dark laying their eggs. Our younger sisters would help feed the babies and keep them clean, while our older sisters would gather food for us.

Life Underground

If we were winter ants, we would not be able to hear well, and it's quiet so far underground anyway. We wouldn't hear children running over us or leaves falling on our entrance. We wouldn't know somebody's dad's car just parked next to our own driveway. Beneath the roots, we wouldn't get wet when the sprinkler showers over our home and across the lawn in the summertime. We wouldn't hear the thud of the family dog flopping right on top of us to gnaw on a tennis ball. But it would all be happening above us, all over the United States. If we were winter ants, we'd miss out on a lot of the fascinating lives of people. We're lucky we're not winter ants. We're people, active all year long, and able to understand and delight in the winter ant's secret wonderland, deep below our feet.

FREQUENTLY ASKED QUESTIONS

Ant researchers often get asked some really interesting questions. For example, what do you do when you get ants in your pants? (Jump around and squeal, obviously!) We like that people pay attention to what's living around them and want to know more. I asked some of my favorite ant researchers what they most often are asked about ants. Here are the answers to some of our most commonly asked ant questions.

What is the biggest ant?

The dinosaur ant, *Dinoponera gigantea*, boasts the largest workers in the world, measuring a little over an inch to more than 1.5 inches long. These whoppers live in South America. But the contest is a close one. At 1.2 inches long, both the Southeast Asian giant forest ant, *Camponotus gigas*, a close relative of our US carpenter ant, and the bullet ant, *Paraponera clavata*, from Central and South America, give dinosaur ants a run for their money. In the United States, carpenter ants can measure up to a half-inch long and probably hold the distinction of the largest ants in the country.

What is the fastest ant?

No one has calculated which ant species runs the fastest, though you might judge your local species for yourself in your backyard with some ant bait and a stopwatch. Still, some ants are known to have some pretty quick moves. One type of trap-jaw ant, *Odontomachus* spp., have jaws that shut at lightning speeds—up to 145 miles per hour. They use these quick reflexes to chomp down on prey and threats alike, but they also use their snappy mouthparts to help them battle intruders or quickly exit any scene. When danger rears its ugly head, a trap-jaw ant can either bounce the botherer off its snapping jaws, tossing the intruder aside, or bounce itself off the intruder, launching itself far from danger. Trap-jaw ants can also snap their jaws down on the ground to catapult themselves into the air and away from danger.

How do I find ants?

As you probably already know, ants are super easy to find. They're all around you! It's best to look for ants on warmer days with low wind speeds, but you can look for ants any time you want.

One of the best ways to find them is to coax them into coming out into the open by offering them a snack. Some ant species love sugar while others prefer protein, so read a little about which species you're trying to find before preparing a meal for them. If you just want to see who's around, you can crumble some pecan cookies onto an index card, a piece of paper or tin foil, or any other flattish surface and wait to see who comes to the party. Pecan cookies offer protein and sugar at the same time. Many ant researchers also use canned tuna fish in oil to find their protein lovers and honey or jelly, like apple jelly, mixed with a tiny bit of water for their sugar lovers. Some researchers mix them both together to make a stinky salad that equates to many ants' dream meal.

While you're watching your ant baits, see if you can identify which species show up. Watch to see how they interact at the baits. Some species bicker with each other, while others ignore everybody completely and stick to the task at hand. Some species lay chemical trails back to their nest to get their sisters excited about this new grocery store in town, while others will carry as much home as they can in their mandibles. Still others will run away, only to return carrying their ready-to-help sisters, whom they plop down on the food. You can follow all these ants back to their nests if you have keen eyes.

Another way to look for ants is to look under things. Logs, flower pots, stones, and mulch are all great objects to peek under or pick through. I like to carefully peel back the bark on rotting wood to see who's living there, and I'm rarely disappointed. Just keep your eyes open for insects and other creatures you might not want to find, like snakes and spiders. They like living under bark and logs as much as ants do. It can be exciting to find them, but sometimes they'll startle you. Remember to replace the log, rock, or bark when you are done peeking, though, so the ants and other small wildlife can return to their business after you amble away.

If you want to take ant hunting to the next level, you can set up pitfall traps. For a pitfall trap, you'll need a little liquid dish detergent (or other liquid soap) mixed with some water (and a little rubbing alcohol, if you have it—or, even better, ethanol) and a small, disposable plastic cup. Dig a cup-size hole in the ground and place your cup in it so the lip of the cup lines up with the top of your hole. Try to get a cup with a small mouth on it so that big stuff won't fall in. Then, fill your cup halfway with your detergent mixture. Leave it in the ground for a day or so, and when you come back, you can see who fell in. You can dump your cup's contents into a white-bottomed tray or, if you don't have that, a clear dish with a white piece of paper under it so you can see everybody. If you leave the cup in the soil too long, the insects you catch will begin to rot and the smell will probably prevent you from wanting to check, so it's best to just leave it out for a day at a time.

If you want to get even deeper into ant-finding, visit your local library and check out a great ant-hunting method book like *Ants: Standard Methods for Measuring and Monitoring Biodiversity*, edited by Donat Agosti, Jonathan Majer, Leeanne Alonso, and Ted Schultz. Books like this will give you all sorts of ideas about how to find the ants you want. Such books are written for serious ant enthusiasts, who, as we all know, include kids, bankers, teachers, and anyone else who wants to seriously study the other societies all around us.

Are ants related to termites?

Ants *are* related to termites in that they're both insects, but they're not closely related to termites, and they have very different lifestyles. While ants have queens who store lots of sperm and lay eggs, termites have kings and queens who mate repeatedly over the course of the colony's life cycle.

Ants also develop by means of "complete metamorphosis," which means the eggs hatch into baby ants (larvae) that look more like fly maggots than they do ants. These larvae must pupate (often by spinning cocoons, though some ants forgo this complexity) and undergo metamorphosis before they are adult workers who look like the ants we see on our sidewalks and trees. Baby termites, on the other hand, hatch from eggs looking like miniature termites.

To tell ants apart from termites just by looking, check their waists and antennae. Ants have narrow "wasp waists," while termites look chubbier around the middle. Ants also have elbowed antennae, whereas termite antennae stick straight out.

Are ants related to wasps?

Ants are more closely related to wasps and bees than they are to termites (or most other insects). Ants, bees, and wasps are all members of the scientific order *Hymenoptera*, which means they share both physical characteristics and common ancestors.

How long do ants live?

We're not sure how long many species of ant workers live. Their lifespans often depend on the season (and whether they get stepped on, licked up by an anteater, chomped by a bird, or taken over by a zombie parasite), but they can live anywhere from a couple of months to several years. Ant researchers typically record how long ants live by how long the colony's queen can survive, since she is the colony's beating heart.

Some ant queens live long, regal lives (if a life consisting mostly of waiting to be fed and laying eggs is regal). Others don't. For example, red imported fire ant queens (*Solenopsis invicta*) can live almost seven years, depending on where they live. Winnow ant (*Aphaenogaster* spp.) queens can live up to 13 years. Some *Lasius* queens have been documented as living nearly 30 years. Other ant queens last less than a season.

Do carpenter ants eat wood?

While carpenter ants gnaw wood to build their nests, they don't actually eat it. Like humans, they just whittle while they work.

What does an ant queen look like?

Some ant queens, like Asian needle ant queens, look a lot like their workers. Others are the same color as their workers but have tremendous gasters, good for egg laying. Ant queens usually have larger eyes and bulkier thoraxes than their workers. The ant queens of some species, such as army ants, are so large and different from their workers that they look like another species entirely.

How many different ant species are there?

So far, we know of about 15,000 ant species roaming the earth, and nearly 1,000 of those species call the United States home. Nearly half of these species have not yet been named and more remain undiscovered, which is to say new ant species may well be lurking in your backyard.

HOW TO KEEP ANTS AT HOME

So You Want to Keep Ants as Pets . . .

Acrobat ants walking tightropes, carpenter ants having some serious conversations—now that you've read about all these creatures crawling around you, you might be tempted to capture some to keep around the house. Watching ants go about their business from the comfort of your home can be fun, especially when they're going about their business in a container and not just wandering around your kitchen counter.

Ant Sleuthing

Before you decide to take the leap into ant ownership, consider becoming a grade-A ant sleuth instead. Ant stalking can offer endless fun, and you'll get the chance to observe many more species

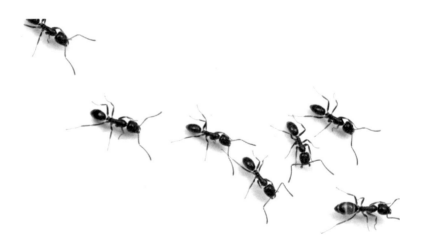

than you would if you kept a species or two in your house. To be a good ant detective, you can either lure the ants to you or hunt them down.

Luring Ants: The Way to an Ant's Heart . . .

When luring ants, consider their favorite meal. Some ants prefer sugary foods (like honeydew), some prefer protein (like arthropods), and some like a little bit of both. Start by reading about your favorite ant species to see what they like best.

Once you know their favorite fare, pick something you have around the house that offers them a taste of what they want. Protein lovers might like a little peanut butter or tuna fish in oil. Those

with a sweet tooth may prefer honey, jam, or sugar water. Cookies are always a good bet. Stay away from "healthy snacks" like broccoli or celery. Ants never seem to care too much for plants (unless those plants have bugs living on them). Feel free to be an ant chef and experiment with different foods as you get to know different ant species.

You'll want a handful of mini platters so you can spread the bait out around your yard. You can use index cards, tin foil squares, or some other flat surface. Place a tiny amount (1/4 teaspoon or less) of your food on each one, and take them outdoors to see what you can find. After you've distributed their bait, wait 20 minutes to an hour and see who shows up.

Watching ants can be very exciting. Some get to the party really early and run away when other ants come around. Others saunter in and frighten away the ants already there. Some ants play well with others and ignore whoever else shows up. Others fight or engage in ritualized displays (where they wave their gasters in the air or open their mandibles or swing their heads to and fro) to show how mean they can be. I've even seen ants show up only to eat the other ants eating the baits!

See how the ants find the baits and how they bring their sisters back to get more food. Do they wander in, or do they make a direct line? Do they come alone, or do they bring reinforcements? If you place two different types of food out for the ants, do they prefer one over the other? You can come up with many questions while luring ants and get a lot of insight into how ants operate just by watching them eat. Be sure not to breathe on them while you watch them. They hate that.

Hunting Ants: No Weapons Necessary

If you've set up a bait, you have a good eye and are quick, you can follow the ants back to their nests. Many of us are not so quick, though, and we need to come up with other ways to hunt ants.

Hunting ants can seem tricky at first, but really it's not so hard. In the same way you figured out what they like to eat, you can figure out where they like to live. Again, start by doing your research. Pick a species or two that live in your area and read about where they like to make their homes. For example, odorous house ants like to live in mulch, and winnow ants like to live in rotting logs.

Next, carefully inspect those areas. On ant hunts, I carry a small but sturdy garden spade to help with delicate digging and exploring. With winnow ants, for example, I like to go into the woods and carefully peel back the bark on rotting logs. It doesn't take long before I find a winnow ant colony in there, running about with their brood in their mandibles, trying to get away from me. I find many other species that way, like acrobat ants, carpenter ants, field ants, and citronella ants. In mulch around the bases of trees or next to houses, I often find Argentine ants or odorous house ants, as well as the occasional tiny thief ant or field ant colony. Sometimes I collect a bunch of acorns and crack them open to see who's living in there. In addition to acorn ants, I'll sometimes find Asian needle ants or odorous house ants. If you have ants living in your home, try to hunt them down by following their trails until you find a nest.

Be careful while you hunt for ants. Know which ants sting and always keep an eye out for other animals that could hurt you—like snakes, centipedes, or even raccoons—that might not want you poking around their homes. More than once, I've plunged my spade in a hollow log without looking only to be greeted by a cohort of angry yellow jackets, ready to tell me they don't like what I've done to their house. If I had paid closer attention, I would have noticed the wasps coming and going from their home and avoided a painful situation for me and my ant-hunting friends. Not paying attention can make you quite unpopular with your companions.

Keeping Ants as Pets

If watching ants in their natural environment isn't enough, and you still want to try your hand at keeping them at home, you have a number of ant-keeping options at your disposal. To get a better idea of what sort of housing might be appropriate, think about the type of ant you'd like to have and its preferred lifestyle. Ant houses work well for species that thrive in enclosed environments, but not others that like to forage over larger areas or up and down tree trunks.

Selecting Your Ants

As you know from reading this book, the world around you is bursting with ant species. Each has its own unique way of life, including diet and housing. You can purchase ants online from one of many ant retailers, but be warned that most retailers send several workers and no queen. Without the queen, the colony will not be able to replenish workers when they die, and the colony will die out in a few months to a year or so. Still, purchasing workers online can be an easy ant-acquiring solution and a rewarding experience for beginner ant keepers.

If you choose to capture your own ants, consider:

× which ant species live around you and are active when you want to collect them

× where those species live (in logs, underground, in mulch, under rocks, high in trees)

× what those species eat (some diets may be more difficult to reproduce at home than others)

× if those species are dangerous and put you or your household at risk. (You will probably want to avoid collecting Asian needle ants, for example, as they sting.)

× if those species could become household pests. (When I studied ants in a laboratory, I accidentally infested my laboratory with Argentine ants more than once. Others do not appreciate this.)

× if the species you select are invasive. Releasing invasive species into the environment, even accidentally, is illegal and can be harmful to the environment

Now that you've decided who you want to bring home, it's time to build or purchase your species' dream house. We'll collect the ants later, when we're ready.

Home, Sweet Home

Ant houses, called formicaria, can be purchased from a number of online retailers. Formicaria constructed from natural materials and customized for a particular ant species work best to ensure you'll be able to keep your ants for a long time. For those just getting started, I'm a fan of purchasing a good formicarium over building one at home. That way, somebody will mail to your house everything your ants need to get rolling, and you can see what works for others before you try it yourself.

When building an ant house, consider how the ant lives outdoors.

If you do decide to build your own ant house, remember that many species can escape containers easily, so you need to have containers with tight-fitting lids. These lids should be fitted with fine screens or have tiny ant-proof holes for proper ventilation.

Formicaria can take many different, beautiful forms.

Ants also need proper moisture. The simplest ant nest can be built using a test tube, a stopper with a hole in it large enough for the ants to crawl out, cotton, tin foil, and water.

Fill the test tube about three-quarters of the way with water. Stuff cotton in the tube with clean fingers until you feel the cotton moisten to your fingertip. You don't want the tube to leak, but you do want it to be moist. Then, put the stopper in the tube. Be sure there's enough room in the tube for your ants to crawl around. Wrap the tube in tin foil so the ants can have a dark environment. You can peel the foil back to observe your colony from time to time. This tube should be placed in your container with tight-fitting lid.

You can feed your ants by placing their food directly in the container. Check your test tube every week or so to ensure it stays moist. If it dries, add another test-tube nest. The ants will move in when they run out of water, and then you can remove their old nest.

Remember, ants do not like disturbances, so with a homemade ant house like this, you should only check on them once or twice a day at most. When you check on them, try to avoid breathing on them, as the carbon dioxide in your breath alarms them.

After you master this basic ant setup, you may want to modify your homemade formicarium to impress your friends, suit your ant-watching desires, or help your ants to live in a more "natural" environment.

If you want to build a home where you can watch your ants crawl around all day, check the Internet for some sample formicaria plans, which will provide you with step-by-step instructions on how to build your ant's dream home. You can also play around with expanding upon your existing setup.

Go Get Your Ants!

Although I do recommend purchasing a premade formicarium, I also recommend collecting the ants for that formicarium yourself. When you collect your own ants, you can get as many as you want, and you can ensure you have a mated queen that could potentially provide you with years of ant-watching pleasure.

The simplest way to collect ants is to go on an ant hunt, outlined above, armed with your spade and containers with tight-fitting lids. When you find the species you want, try to locate the queen, some workers, and a bit of brood. Queens are usually larger than the other ants in the nests and can often (but not always) be found near the brood. Delicately scoop up the queen, workers, and brood with a spade and place them in your container with a tight-fitting lid.

When you get home, place your test-tube nest in that container and wait for the queen and workers to move in. Leave them alone for a day or two so they feel comfortable.

Supper Time!

Now that you have a small colony, it's time to give them something to eat. Most ant species need protein, sugar, fats, and water. Read up on your species to see what types of food they like most. As you work to make fatties, keep in mind that just because they prefer sugary foods doesn't mean they don't need a little protein and fat every now and then. You can purchase ant food online, or you can give it a go yourself. Here is a list of foods I use:

PROTEIN
× Dead arthropods (sized appropriately for your species)
× Peanut butter (works sometimes but not all times)
× Seeds and nuts (for ants interested in seeds and nuts)
× Tuna fish in oil (they love it, but it smells bad)
× Pecan cookies (makes a nice snack but will not sustain the colony)

SUGAR
× Sugar water (50 percent sugar, 50 percent water)
× Honey mixed with a little water
× Apple jelly or some other fruit jelly mixed with a little water

FAT
× Tuna in oil has fats, as do seeds and peanut butter (above)
× Some people mix a little olive or canola oil with their sugar solutions

With all of these foods, it's important to remember ants' tiny stomachs. Because you want to avoid getting mold in your ant houses, try not to feed them too much, and remove uneaten food as it dries out or begins to mold. I like to feed them on tiny trays I fashion out of tin foil or in shallow bottle caps, so I can remove uneaten food easily.

Watching Your Ants

Now that you have ants in your living room, spend some time watching them to see what they do. How do they eat? When you put out new food for them, how do they find it? How long does it take them to find it? Which foods do they pick up most often and most readily? Which foods do they avoid? What do they do in their nest chambers all day? Does light bother them? What times of year does the queen lay eggs? How do they feed their larvae? Watching ants can be fascinating. You'll be amazed how quickly the time goes.

Keeping ants at home can be complicated. Mack Pridgen, founder of the ant house business Tar Heel Ants, has figured out some valuable tricks to the trade, and he wants to share them with you.

DR. ELEANOR: How did you get interested in keeping ants?

MACK: I was fascinated by ants as a kid. We had some very large colonies of carpenter ants (*Camponotus*) around my house, and I was the kid with the jars trying to collect them and feed them. My favorites were the giant red ants with big heads! I now know those as the carpenter ant *Camponotus castaneus* majors.

Years later I bought my daughter a gel ant farm as a Christmas gift and decided to show her how cool ants were to watch digging. Instead of ordering ants (when I was a kid, mine arrived dead), I went out the day after Christmas and found carpenter ant *Camponotus chromaiodes* workers foraging up a tree. They died within a couple of days. I pressed on, researching how people cultivate ants in labs, and I reached out to the local ant lab here in town. Before long I was raising young colonies in test tubes and planning their first formicarium!

DR. ELEANOR: What common ants would you say are the "best" (easiest, safest) for beginners to keep at home?

MACK: It depends on where you live. Concentrated in the southeastern United States but also found in other regions throughout the country, various winnow ants are abundant in forests under rocks. Winnow ants are always my first suggestion. I was lucky enough to have them for my first large colony.

Carpenter ants (*Camponotus* spp.) also work well for beginners. Their queens are huge, around three-quarters of an inch in size, and they vary in color.

DR. ELEANOR: What do you find to be some of the common mistakes people make when keeping ants at home?

MACK: Common mistakes include

1. **Improper feeding expectations and procedures.** For example, most queens do not need food until their first workers have emerged. The workers find food and feed the queen. Many inexperienced ant keepers try to feed ants live insects, too much of a dead insect, or sweet liquids inside their claustral chambers (test tubes or other starting formicaria). This stresses the queen, causes mold problems for the queen and her brood, and more. Carefully feeding small amounts of foods is OK but the excess should be removed immediately. Practicing a proper feeding regimen can make all the difference for your colony, and an improper regimen is often an ant keeper's biggest problem. Follow these proper feeding techniques:

   × Offer small pieces of fresh insects often—at least a couple times a week—and anchor them to something (cork, silicon, etc.) to prevent trash from entering the formicarium (less important if you can disassemble your formicarium).
   × Always provide liquids and solids on dishes, regardless of colony

size. Any food that smears on the formicarium surfaces or foraging area can mold and lead to major problems for the colony down the road. Providing some loose (ant-safe) substrate allows your ants to cover the problem areas themselves.

× Never feed your colonies live food. Workers can get injured battling living insects. At minimum, injure prey if live feeding is a must for your particular species.

× Rotate foods. Use a combination of fresh fruits (apples are a personal favorite), dead crickets, mealworms, and fruit flies, in addition to other foods that provide a good balance of carbohydrates, proteins, and fats.

2. **Not anticipating how easily ants can climb and escape.** One of the first major considerations while planning your homemade formicarium is security. Think of this as protection for your ants, not you!

3. **Not planning ahead.** Colonies can grow fast during their growth season. Many of the species we see in urban areas, such as pavement ants, odorous house ants, and acrobat ants, can quickly outgrow their formicaria and can be difficult to contain and feed if not kept in a proper-size habitat.

DR. ELEANOR: What are your top tips for building an ant house at home?
MACK:

1. **Use what you already know.** You can build an ant home using skills you already have. Good with crafts? Perhaps build a plastic nest out of acrylic or make a plaster nest. Artistically inclined and like to carve? Good with power tools? Try finding some Ytong (AAC block) bricks and crafting your own design.

2. **Plan your hydration first.** Moisture is your colony's lifeline. If you can, use a hygrometer to test the moisture level in your formicarium before introducing your ants.

3. **Ask questions.** Many people have been where you are now. Join an online group or forum.

4. **Keep local ants!** Don't risk spreading harmful species.

DR. ELEANOR: What are your top five tips for keeping ants alive at home?

MACK:

1. **Monitor the temperature** near your colonies daily with a thermometer. Keep your ants at a constant temperature (78–82°F, typically, though some species may prefer warmer conditions). Quick temperature changes can be problematic for ants. Remember: Ants are not plants. They don't do well in the sun!

2. **Label your ant colonies** in case friends or families don't know what they are. Shaking or tilting your formicarium is not recommended.

3. **Wear disposable latex gloves** or, at a minimum, wash your hands before feeding your ants or handling their formicaria.

4. **Use organic fruits and honey** when possible.

5. **Clean out your ant foraging areas regularly**, whenever you see detritus build up.

EPILOGUE:
THE VALUE OF
OUR COLLECTIONS,
AND YOURS

DR. CORRIE S. MOREAU AND DR. EDWARD O. WILSON

Ants are everywhere. You can find them in the rainforests of the tropics, in the woods behind your house, or even in the cracks of the sidewalk in cities. This means that anyone, anywhere (except, surprisingly with such a seemingly appropriate name, Antarctica) can find and observe ants. Not only are they often locally abundant and species diverse, but as social animals they carry out often quite complex tasks and behaviors, making them one of the best groups of animals to observe and study. Not only can you watch them cooperate to drag off a dead cricket or cookie crumb, but you can also collect all the individuals you find to determine how many species are living in your backyard.

This is precisely what drew us to lifelong careers studying ants. Coincidentally we both grew up in the southern United States and spent many hours outside collecting and observing nature. We kept ant farms on our desks, jars of bugs under our beds, and lizards and snakes in our rooms. This collecting is important not only for advancing scientific knowledge but also for sparking and maintaining human interest in the natural world. Many of you have probably experienced this firsthand through your own personal collections or by participating in the School of Ants project (schoolofants.org), which helped make the species lists in this book possible. Although we are less likely to keep live ants under our beds today, we do still indulge in this passion for collecting in order to conduct scientific research and add to our knowledge of nature.

Collecting animals, plants, fungi, and even bacteria is important for describing, researching, and conserving biological diversity (also called biodiversity), but equally important is making sure those individual collected specimens or species are preserved through time. This is where natural history museums come into play. Natural history museums are not places to read about historical reenactments, as the name may suggest to some. They are dynamic places where visitors can experience the natural world through museum exhibitions and where active research by scientists from around the world continues.

Natural history museums are time machines. Their collections are windows into evolution. Scientists can use museum collections to study the past, document the diversity of life on the planet today, and predict what biodiversity will look like in the future. Natural history museums have many, many more specimens and artifacts behind the scenes than are on display. For example, the Field Museum of Natural History in Chicago has more than 26 million specimens and artifacts, but less than 1 percent are on display in the public areas. The same is true of the Harvard Museum of Comparative Zoology in Cambridge, which houses over 21 million specimens. But the specimens not on exhibit serve another important purpose. Natural history museums are also scientific species libraries. The specimens and artifacts in these vast collections are used by scientists the world over, who can visit these museums to study and research the collections. Or, these specimens can be loaned to scientists anywhere in the world to allow them to complete their research.

These scientific collections are not static in time and must continue to grow. Not only do they ensure that we have representatives of all species found on the planet, from the past (fossils) to the present, but these collections are also the foundation for understanding where species are found. Imagine if only one individual monarch butterfly was ever collected and put in a museum. You might logically conclude that this species can only be found in that one location. But if we include representatives of that species from across all the locations it can be found, from Canada to South America, you would have a much more complete picture of where to expect to find monarch butterflies and understand how they fit in the larger ecosystem. Gathering specimens and information about where they can be found is also important to understand pest species. In addition, these warehouses of biological diversity permit scientists to address questions about the genetic diversity of living and extinct organisms and provide data to help protect and conserve biodiversity the world over.

Our wish for you is to continue to observe and study the ants (and other species) in your back yard. There are still many, many unanswered questions to be addressed, scientific puzzles to be solved, and species in need of champions to fight for their protection.

Córrie S. Moreau
Field Museum of Natural History
Department of Science and Education
Chicago, Illinois

Edward O. Wilson
Harvard University
Museum of Comparative Zoology
Cambridge, Massachusetts

ACKNOWLEDGMENTS

This book would not have been possible without the expertise and help of many great individuals. Thank you, Rob Dunn and Andrea Lucky, for having the curiosity and vision to help us all turn over stones and peer in the grass to find the ants around us. Rob, thank you, also, for your editorial guidance, opportunities, and encouragement. You and the University of Chicago Press have opened the door for us to share our joy of ants. My great appreciation goes to the University of Chicago Press's Christie Henry, Gina Wadas, Logan Ryan Smith, and Amy Krynak for their guidance and support. Robin Anders, you know how to edit and critique like nobody else. Thank you. Thank you also, Kathryn and Jamesie Spicer, for your editorial assistance and for giving me the crumbs and long mornings and afternoons to meet my ants. Neil McCoy, thank you for your creativity, which helps to sharpen and enliven ideas. Holly Menninger, your powers of coordination and organization are unparalleled. Thank you, Alex Wild, for using your skill and vision to make giants of ants, for showing us how beautiful their tiny world is and how they can be our friends. Thank you, E. O. Wilson and Corrie Moreau, for your words and your discoveries. Thank you for your expert advice and encouragement, Matt Shipman. Russ Campbell and the Burroughs Wellcome Fund, thank you for recognizing this book and giving it a chance. Thank you to my ant people, including Jules Silverman, Alexei Rowles, John Brightwell, Bill Reynolds, Sean Menke, Jon Shik, Brad Powell, David Bednar, Grzesiek Buczkowski, Heike Meissner, Benoit Guenard, Clint Penick, and Amy Savage, plus one snake guy, Warren Booth. Greg Rice, you are the words and the ants and everything else. Thank you.

GLOSSARY

abdomen: The third major division of the insect body (aka rump, booty, posterior, etc.) that contains most of the ant's organs and its stinger.

ant: A small, wingless, wasp-like insect that usually lives in eusocial groups. An incredibly diverse and ecologically important animal. Earth houses more than 15,000 named ant species, and many more awaiting scientific description and naming. *See* Hymenoptera and Formicidae.

antenna (pl. antennae): A segmented appendage projecting from either side of an adult insect's head. Antennae function as sensory organs and help ants sniff, feel, and taste.

aphid: A small, plant fluid–sucking insect that usually resembles a tiny cicada or a tiny, chubby katydid. Aphids can be winged or wingless and usually are found on the undersides of plant leaves or along stems. Often protected by many ant species, aphids turn excess plant fluid into a sweet substance called honeydew, which ants eat.

arthropod: An animal with jointed legs and an exoskeleton. Arthropoda refers to the large scientific group including shellfish, insects, scorpions, and spiders. *Arthro* comes from the Greek word meaning "joint," and *poda* comes from the Greek word meaning "foot." The vast majority of described species on earth are arthropods. Ants are arthropods.

biodiversity: The amount of different life forms in an area. In general, a rich biodiversity (lots of different life forms) means a healthy environment. Some invasive ant species, like Asian needle ants, reduce biodiversity when they move into an area, which could result in an unhealthy habitat.

caste: Refers to the various groups of ants within a colony. Sexual castes consist of two groups: males and females. Morphological castes consist of two or more groups, typically minors and majors (soldiers). Temporal castes divide ants according to age and the jobs they do at those ages. Reproductive castes refer to queens, which reproduce, and workers, which don't.

colony: A group of ants, often closely genetically related, which operate as a functional unit without aggression among the group. Colonies can have one or many nests and one or many queens.

common name: The moniker we call ants for convenience. Most people referring to ants use their common names. Common names usually refer to some aspect of the ant's appearance (like "little black ant") or behavior (like "fire ant"). Common names can be different in different languages.

complete metamorphosis: A form of insect development, in which the insect undergoes the following stages to adulthood: egg larva (looks very different from adults) pupa adult. Ants undergo complete metamorphosis.

crop: A "stomach" attached to the esophagus that serves to receive and hold food. It's like an internal backpack. Crops hold food without digesting it so ants can share it with their sisters or eat it later.

ecology: The study of the relationships between living things and their environment. *Eco* comes from the Greek word for "house," and *ology* comes from the Greek word meaning "the study of."

egg: The first stage in an ant's development and laid by queens, an egg has a simple germ cell, nutritious yolk, and a surrounding membrane.

entomology: The study of insects and other arthropods. *Entom* comes from the Greek word for "insect," and *ology* comes from the Greek word meaning "the science of."

entomologist: Someone who studies insects and other arthropods.

eusocial: If an animal cooperatively cares for its young, has a reproductive division of labor (for example, queens reproduce; workers work), and an overlap of at least two generations sharing a space and contributing to the group, then it's eusocial. Most ant species are eusocial. The few that are not eusocial are workerless parasites in the nests of other ants.

exoskeleton: The "hard outer shell" of insects and other arthropods. Instead of bones, insects have a suit of armor consisting of a plastic-like substance called chitin, which is covered by a thin layer of waxy material. Ant muscles are attached on the inside of the exoskeleton.

exotic: In invasion ecology, exotic refers to an organism that is present in an area but which comes from a different place. That is, that organism did not evolve in that area. Exotic species are not always invasive, and they're not always pests. Your pet cat is an exotic species, and some would say it is a pest. Honey bees are also exotic species in the United States. They come from Europe and Africa.

Formicidae: The scientific grouping called family to which all ants belong. The word *Formicidae* comes from the Latin word meaning "ant."

gaster: The swollen part of the abdomen behind the ant's skinny waist, or petiole.

genus (pl. genera): A group of species that share characteristics and are often closely related. For example, thief ants and red imported fire ants share many physical characteristics and are closely related. They share the genus *Solenopsis.* Knowing genera can help you mentally group ants by form and function.

holometabolous: The quality of an organism, like an ant, of undergoing complete metamorphosis.

honeydew: A sugary fluid excreted from the abdomens of many different insects, including aphids and scale insects. Many ant species love to eat honeydew and rely on it for survival.

Hymenoptera: The scientific order of insects to which ants belong. Bees and wasps also belong to this order, and these three types of insects share much in common, including their skinny waists and their tendency toward forming social groups. Hymenoptera comes from *hymen*, the Greek word for "membrane," and *ptera*, the Greek word for "wings." Hymenopteran wings look like a thin membrane stretched across a few veins.

insect: A class of animal that has an exoskeleton, three major body segments (head, thorax, and abdomen), six legs, and two antennae. Ants are insects. Spiders (eight legs, two segments) are not.

invasive species: A species that moves into an area and negatively impacts that environment. Red imported fire ants and Asian needle ants are examples of invasive species.

invertebrate: A general term referring to any animal that does not have a backbone. Worms, insects, crabs, octopi, and spiders are all invertebrates. Most of life on earth has no backbone.

larva (pl. larvae): The second stage in an ant's development, between egg and pupa. Larvae differ in form from adults. Ant larvae often look like legless grubs.

major: A worker subcaste (*see morph*) in which the individual is typically larger and specialized for defense. Big headed ants have the most prominent majors, but other ant species, like carpenter ants, can have majors, too. Sometimes referred to as soldiers.

mandible: The first pair of jaws in ants. Mandibles usually stick out from the front of the head and are good for chomping, slicing, and carrying.

minor: A worker subcaste (*see morph*) in which the individual is typically smaller and specialized for work.

morph: Any of the various forms of ants within a caste. For example, a major is one morph, while a minor is another.

myrmecologist: A person who studies ants.

native species: An organism that is present in an environment "naturally" and not because a human facilitated its presence in the environment.

nest: Among ant species, a nest is a discrete living space for a related group, usually containing workers, brood, and queens, but sometimes containing any two of the three (or any combination plus males). Nests can be as simple as a hangout spot under a rock (as with odorous house ants) or as complex as intricate underground tunnel networks connecting rooms (as with winter ants). Ant species can have one or many nests per colony.

nestmate: Individuals, usually related, who share a nest. Nestmate can also refer to members of the same multinest colony who don't share a particular nest. Nestmates do not fight one another when they meet outside the nest, and they recognize one another as nestmates because they smell alike.

pest: A species that negatively impacts its environment. Some ant pests, like odorous house ants, are "nuisance pests," meaning that people tend to be bothered by them but they don't necessarily negatively impact the environment. The primary damage associated with these pests is indirect, resulting from people's use of hazardous chemicals to exterminate them. Other pest ants, like Asian needle ants, are true pests, meaning they cause economic damage (like crop loss), or pose health risks (as from the sting of an Asian needle ant). Not all pests are invasive species or exotic species, and not all exotic species are pests.

petiole: The skinny segments at the beginning of the abdomen, between the thorax and gaster, that give ants their skinny, wasp-like "waists."

pheromone: Any one of many chemical secretions used to communicate within species. Ants use a variety of pheromones to communicate, including alarm pheromones, recognition pheromones, and trail pheromones.

polyethism: The division of labor among members in the colony. Different forms of polyethism are apparent in ant colonies. For example, many ants display something called age-based polyethism, where younger workers perform different tasks than older workers.

polymorphism: In ants, having several physical forms of adults. Many insects display polymorphism.

pupa (pl. pupae): The life cycle stage in insects with complete meta-morphosis. In ants, it occurs between the larva and adult stages, when the insect becomes inactive, doesn't eat, and develops the physical features of an adult.

queen: In ants, female colony members who can lay fertilized eggs. Usually larger than workers.

scale insect: A small, plant fluid–feeding insect that looks like a bump, shell, or scale stuck to plant bark or stems. Often protected by many ant species, scale insects turn excess plant fluid into a sweet substance called honeydew, which ants eat.

scientific name: The formal epithet used to describe species; regulated by a huge international formal naming process. Usually with Greek or Latin roots, scientific names are the same in all languages across the globe. This standardization is extremely useful for communicating science. Just like we have first and last names, scientific names consist of two parts: one for genus and the other for species. As knowing somebody's name can tell you about that person, knowing scientific names can tell you a lot about the insect. The genus name is like our last name and the species name is like our first name. For example, my name is Eleanor Spicer Rice. If I told somebody from my hometown my name, that person would know I'm kin to the Spicers and could have a general knowledge about me before they even got to know me. If she knew my relatives, she could get an idea of what I might look like and could have an idea of where I live and to a certain degree how I might behave. If you tell an ant scientist you saw a *Brachyponera chinensis*, even if he's never met one, he would know a lot about how the species looks, lives, and acts if he knows other *Brachyponera*. Just as my first name, Eleanor, distinguishes me from the other Spicers hanging around town, the specific epithet distinguishes each species from all the other species and gives us an idea of what that species does. For our *Brachyponera chinensis*, "*chinensis*" tells us this species is native to Asia. While species might have the same specific name, no two species share both genus and species name. That way, there's no confusion about which species scientists are talking about.

segment: In insects, any division of the body. While segment can refer to each joint in the leg or antenna, we most often think of segments when discussing one of the three major insect body divisions: head, thorax, and abdomen.

soldier: See major.

species: A group of individuals that are genetically similar and able to mate and produce offspring that can also mate and produce offspring.

spiracle: The holes on an insect's body that open to its respiratory, or tracheal, system. Basically, it's how the insect breathes. Like our mouth or nose.

thorax: The second, or middle, segment of an insect. The thorax is the locomotion center.

trophallaxis: In ants and other eusocial insects, the process of exchanging crop contents between individuals through the mouth. It's one way ants share food and communicate information.

worker: In social insects like ants, a member of the laboring caste that isn't able to reproduce.

ADDITIONAL RESOURCES

Throughout the book, we relied on a collection of excellent resources that we hope will help you, too, as you continue your love affair with ants.

WEBSITES

- × Antweb (www.antweb.org) offers the world's largest ant-centric database, complete with photographs, cool ant stats, and the latest research from curators across the country.
- × Alex Wild snaps up-close photos of thousands of ants and displays them on www.alexanderwild.com.
- × Joe MacGown's beautifully illustrated ant keys may help you identify your formicid neighbors. Explore his findings, plus a bunch of other great insect information, on the Mississippi Entomological Museum website (http://mississippientomologicalmuseum.org.msstate.edu).
- × With AntWiki (www.antwiki.org) and AntCat (www.antcat.org), you can check out all the ant taxonomic information you could ever want.
- × And if you ever want to read scientific literature about ants, a great place to start is the USDA's FORMIS page (http://www.ars.usda.gov/News/docs.htm?docid=10003).

BOOKS

- × *Ants of North America: A Guide to the Genera*, by Brian L. Fisher and Stephan P. Cover, is the best ant field guide around. In addition to the helpful and beautiful ant graphics, the authors give you an engaging natural history of each genus, helping you to get to know these ants on a deeper level.
- × *Journey to the Ants. A Story of Scientific Exploration*, by Bert Hölldobler and E. O. Wilson is an exciting, beautiful book about the discovery and love of our natural world (through the eyes of two of the world's best and most treasured ant lovers).

× *Adventures among Ants. A Global Safari with a Cast of Trillions*, by Mark Moffett. The "Indiana Jones of Ants" shares personal stories and interesting ant behaviors around the world in this page turner.

× *The Ants*, by Bert Hölldobler and Edward O. Wilson, is the definitive guide to all things ants. Despite its intimidating size, this book is an engaging read and illuminates every ant nook and cranny, from ant evolution to their array of behaviors to their intricate physiologies and taxonomic details.